Workflow Scheduling on Computing Systems

This book will serve as a guide in understanding workflow scheduling techniques on computing systems such as Cluster, Supercomputers, Grid computing, Cloud computing, Edge computing, Fog computing, and the practical realization of such methods.

It offers a whole new perspective and holistic approach in understanding computing systems' workflow scheduling. Expressing and exposing approaches for various process-centric cloud-based applications give a full coverage of most systems' energy consumption, reliability, resource utilization, cost, and application stochastic computation. By combining theory with application and connecting mathematical concepts and models with their resource management targets, this book will be equally accessible to readers with both Computer Science and Engineering backgrounds.

It will be of great interest to students and professionals alike in the field of computing system design, management, and application. This book will also be beneficial to the general audience and technology enthusiasts who want to expand their knowledge on computer structure.

Kenli Li is currently a full professor of computer science and technology at Hunan University, and the deputy director of National Supercomputing Center in Changsha. His major research areas include parallel computing, high-performance computing, grid, and cloud computing.

Xiaoyong Tang is currently Professor at the School of Computer and Communication Engineering, Changsha University of Science and Technology. He has published more than 50 technique papers in international journals and conferences. He holds over 10 patents announced or authorized by the Chinese National Intellectual Property Administration. He is among the World's Top 2% Scientists.

Jing Mei is currently Assistant Professor at the College of Information Science and Engineering, Hunan Normal University. Her research interests include parallel and distributed computing, cloud computing, edge computing etc.

Longxin Zhang is currently Associate Professor of computer science with the Hunan University of Technology. He is also a Visiting Scholar of the University of Florida.

Wangdong Yang is Professor of computer science and technology at Hunan University, Changsha, China. His research interests include modeling and programming for heterogeneous computing systems and parallel algorithms.

Keqin Li is a SUNY Distinguished Professor of computer science at the State University of New York. He is also National Distinguished Professor with Hunan University, China. He holds over 70 patents announced or authorized by the Chinese National Intellectual Property Administration. He is among the World's Top 5 most influential scientists in distributed computing based on a composite indicator of Scopus citation database.

Workflow Scheduling on Computing Systems

Kenli Li, Xiaoyong Tang, Jing Mei, Longxin Zhang, Wangdong Yang, and Keqin Li

CRC Press
Taylor & Francis Group
Boca Raton London New York

CRC Press is an imprint of the
Taylor & Francis Group, an **informa** business

This book is published with financial support from National Key R&D Programs of China (Grant No. 2020YFB2104000), National Natural Science Foundation of China (Grant No. 61972146), and Hunan Provincial Natural Science Foundation of China (Grant No. 2020JJ4376).

First edition published 2023
by CRC Press
6000 Broken Sound Parkway NW, Suite 300, Boca Raton, FL 33487-2742

and by CRC Press
4 Park Square, Milton Park, Abingdon, Oxon, OX14 4RN

CRC Press is an imprint of Taylor & Francis Group, LLC

Library of Congress Cataloging-in-Publication Data

Names: Li, Kenli, author.
Title: Workflow scheduling on computing systems / Kenli Li, Xiaoyong Tang, Jing Mei, Longxin Zhang, Wangdong Yang, and Keqin Li.
Description: First edition. | Boca Raton, FL : CRC Press, 2023. | Includes bibliographical references. | Summary: "This book will serve as a guide in understanding workflow scheduling techniques on computing systems such as Cluster, Supercomputers, Grid computing, Cloud computing, Edge computing, Fog computing, and the practical realization of such methods. It offers a whole new perspective and holistic approach in understanding computing systems' workflow scheduling. Expressing and exposing the approaches for various process-centric cloud-based applications, which gives a full coverage of most systems' energy consumption, reliability, resource utilization, cost, and application stochastic computation. By combining theory with application and connecting mathematical concepts and models with their resource management targets, this book will be equally accessible to readers with both Computer Science and Engineering backgrounds. It will be of great interest to students and professionals alike in the field of computing system design, management, and application. This book will also be beneficial to the general audience and technology enthusiasts who want to expand their knowledge on computer structure"-- Provided by publisher.
Identifiers: LCCN 2022004960 (print) | LCCN 2022004961 (ebook) | ISBN 9781032309200 (hbk) | ISBN 9781032309217 (pbk) | ISBN 9781003307273 (ebk)
Subjects: LCSH: Workflow management systems.
Classification: LCC HD62.175 .L53 2023 (print) | LCC HD62.175 (ebook) | DDC 005.74--dc23/eng/20220506
LC record available at https://lccn.loc.gov/2022004960
LC ebook record available at https://lccn.loc.gov/2022004961

ISBN: 978-1-032-30920-0 (hbk)
ISBN: 978-1-032-30921-7 (pbk)
ISBN: 978-1-003-30727-3 (ebk)

DOI: 10.1201/b23006

Typeset in LM Roman
by KnowledgeWorks Global Ltd.

Contents

List of Figures

List of Tables

Foreword

In recent years, with the popularity of the Internet and the availability of powerful computers and high-speed networks as low-cost commodity components, it is possible to construct large-scale parallel and distributed computing systems, such as cluster systems, supercomputers, grid computing, cloud computing, and edge/fog computing. These technical opportunities enable the sharing, selection, and aggregation of geographically distributed heterogeneous resources to solve science, engineering, and commerce problems. Resource management plays a key role in improving the performance of these systems, and especially effective and efficient scheduling methods are fundamentally important. However, the systems face a lot of challenging problems, such as energy consumption, reliability, resource utilization, cost, instability, and resource contention. Workflow scheduling aims at meeting user demands and resource provider management indicators, while maintaining a good overall performance or throughput for computing systems. The publication of this book satisfies this need in a timely manner.

This book offers a systematic presentation of workflow scheduling, which encompasses the systems architecture, scheduling model, energy consumption, reliability, resource utilization, problem formulation, billing mechanisms, and the detailed discussion of the theoretical underpinnings, design methodology, and practical implementation. This book is rich in content and detailed in graphics. For each presented algorithm, the book uses corresponding motivational examples to explain clearly and achieve the easy-to-understand purpose. In particular, the book:

- Offers a comprehensive overview of computing systems work-flow scheduling techniques about systems, scheduling architecture, energy consumption, reliability, resource utilization, problem formulation, billing mechanism, methods, design considerations, and practical implementation.

- Presents the design principles necessary for analyzing the computing systems requirements, objectives, time complexity and constraints, that will guide engineering students and engineers toward achieving high-performance, low-cost, and efficient resource management systems.

- Demonstrates the practical implementation of workflow scheduling and their design guidelines and optimizations that can be directly adopted in engineering application and research work.

- Provides a complete perspective on workflow scheduling that hopefully can inspire appreciation and better understanding of the subject matter.

It is a great pleasure to introduce this Workflow Scheduling on Computing Systems, which is a joint effort and creation of six scholars with dedication and distinction. The authors have published very extensively in the fields of grid computing systems, cluster systems, cloud computing, and are undoubtedly the leading scholars in scheduling workflow parallel applications on computing systems. Finally, I would like to congratulate the authors on their excellent work, and I look forward to see the publication of this book.

Kai Hwang
Presidential Chair Professor
Chinese University of Hong Kong
Shenzhen, China

Author Bios

Kenli Li (Senior Member, IEEE) received his PhD in computer science from the Huazhong University of Science and Technology, China, in 2003. He was a visiting scholar at the University of Illinois at Urbana-Champaign, Champaign, Illinois from 2004 to 2005. He is currently a full professor of computer science and technology at Hunan University, China, and deputy director of National Supercomputing Center in Changsha. His major research areas include parallel computing, high-performance computing, grid and cloud computing. He has published more than 130 research papers in international conferences and journals such as the IEEE Transactions on Computers, IEEE Transactions on Parallel and Distributed Systems, IEEE Transactions on Signal Processing, Journal of Parallel and Distributed Computing, ICPP, and CCGrid. He is an outstanding member of CCF. He is serves on the editorial board of the IEEE Transactions on Computers.

Xiaoyong Tang received his M.S. and Ph.D. degrees from Hunan University, China, in 2007 and 2013, respectively. He is currently Professor at the School of Computer and Communication Engineering, Changsha University of Science and Technology. His research interests include parallel computing, cloud computing, big data, modeling and scheduling of distributed computing systems, distributed system reliability, and parallel algorithms. He has published more than 50 technique papers in international journals and conferences. He holds over 10 patents announced or authorized by the Chinese National Intellectual Property Administration. He is among the World's Top 2%

Scientists. He is the reviewers of TC, TPDS, TII, JPDC, FGCS, CPE, JS, and so on.

Jing Mei received her Ph.D in computer science from Hunan University, China, in 2015. She is currently an assistant professor in the College of Information Science and Engineering in Hunan Normal University. Her research interests include parallel and distributed computing, cloud computing, edge computing etc. She has published 16 research articles in international conference and journals, such as IEEE Transactions on Computers, IEEE Transactions on Parallel and Distributed System, IEEE Transactions on Service Computing, Cluster Computing, Journal of Grid Computing, and Journal of Supercomputing.

Longxin Zhang received his Ph.D. in computer science from Hunan University, China, in 2015. He is currently Associate Professor of computer science at the Hunan University of Technology. He is also a Visiting Scholar of the University of Florida. His major research interests include modeling and scheduling for distributed computing systems, distributed system reliability, parallel algorithms, cloud computing, and deep learning. He has published more than 20 technique papers in international journals and conferences. He is the reviewer of IEEE TPDS, ACM TIST,IEEE TII, IEEE IOT, INS, ASC, and so on.

Wangdong Yang received his PhD in computer science and technology from Hunan University, China. He is a professor of computer science and technology in Hunan University, China. His research interests include modeling and programming for heterogeneous computing systems and parallel algorithms. He has published more than 30 papers in international conferences and journals such as the IEEE Transactions on Computers, the IEEE Transactions on Parallel and Distributed Systems, etc.

Keqin Li is a SUNY Distinguished Professor of computer science at the State University of New York. He is also a National Distinguished Professor of Hunan University, China. His current research interests include cloud computing, fog computing and mobile edge computing, energy-efficient computing and communication, embedded systems and cyber-physical systems, heterogeneous computing systems, big data computing, high-performance computing, CPU-GPU hybrid and cooperative computing, computer architectures and systems, computer networking, machine learning, intelligent and soft computing. He has authored or coauthored more than 840 journal articles, book chapters, and refereed conference papers, and has received several best paper awards. He holds over 70 patents announced or authorized by the Chinese National Intellectual Property Administration. He is among the world's top 5 most influential scientists in distributed computing based on a composite indicator of Scopus citation database. He has chaired many international conferences. He is currently an associate editor of the ACM Computing Surveys and the CCF Transactions on High Performance Computing. He has served on the editorial boards of the IEEE Transactions on Parallel and Distributed Systems, the IEEE Transactions on Computers, the IEEE Transactions on Cloud Computing, the IEEE Transactions on Services Computing, and the IEEE Transactions on Sustainable Computing. He is an IEEE Fellow.

Preface

MOTIVATION OF THE BOOK

In the past few years, with the rapid development of IT technology, computing systems have become the core infrastructure of social economy. However, with the exponential growth of computing and data storage requirements, computing systems are facing with a lot of challenging problems, such as energy consumption, reliability, resource utilization, cost, stochastic computation, and resource contention. Workflow scheduling aims at meeting user demands and resource provider management indicators while maintaining a good overall performance or throughput for such systems.

With the increasingly prominent role of workflow scheduling on computing systems, it is timely to introduce the workflow scheduling technology, including the basic concept of workflow scheduling, stochastic tasks scheduling, reliability-driven scheduling, reliability-energy-aware scheduling, interconnection network-aware scheduling, and resource-aware duplication optimization scheduling. To the best of our knowledge, although many books about job or task scheduling already exist, these books lack to provide a comprehensive review and thorough discussion of workflow scheduling. Educating and imparting the holistic understanding of workflow scheduling on computing systems has laid a strong foundation for postgraduate students, research scholars, and practicing engineers in generating and innovating solutions and products for a broad range of applications.

In recognition of this, the book *Workflow Scheduling on Computing Systems* is intended to provide a coverage on the theoretical and practical aspects of the subject matter, which includes not only the conventional workflow scheduling but also the systems challenging problems, such as energy consumption, reliability, resource utilization, cost, and all of which stem from the authors' own research work.

SUMMARY OF CONTENTS

This book focuses on workflow scheduling on computing systems. The main contents are summarized as follows.

Chapter 1 introduces the working principle of resource management and some typical resource managements (such as SLURM, PBS,YARN) in computing systems. Then, this chapter presents the practical application of workflow DAG model and real-world workflow applications.

In Chapter 2, we introduce the scheduling problems, workflow task scheduling, scheduling challenges, and the classification of scheduling algorithms. We also list several typical heuristic workflow scheduling algorithms such as DLS, MCP, HEFT.

Chapter 3 focuses on the stochastic scheduling problem on grid computing systems. In order to effectively scheduling precedence constrained stochastic tasks, this chapter present a stochastic heterogeneous earliest finish time scheduling algorithm, which incorporate the stochastic attribute, such as expected value and variance, of task processing time and edge communication time into scheduling.

Chapter 4 emphasizes the scheduling stochastic parallel applications with precedence constrained tasks on heterogeneous cluster systems. It formulates the stochastic task scheduling model and develops effective methods to deal with the normally distributed random variables. This chapter also describes a stochastic dynamic level scheduling algorithm SDLS, which employs stochastic bottom level and stochastic dynamic level to produce schedules of high quality.

In Chapter 5, we first build a reliability and energy-aware task scheduling architecture including precedence-constrained parallel applications, energy consumption model on heterogeneous systems. Then, we present the single processor failure rate model based on Dynamic Voltage and Frequency Adjustment (DVFS) technique and deduce the application reliability of systems. Finally, to provide an optimum solution for this problem, a heuristic reliability-energy aware scheduling algorithm is presented.

Chapter 6 addresses a bi-objective genetic algorithm to deal with the bi-objective optimization problem of high system reliability and low energy consumption for parallel tasks. This approach offers users more flexibility when jobs are submitted to a data center.

Chapter 7 comprehensively presents the issues of heterogeneous systems, energy consumption of processors and interconnection networks, computation-intensive scientific workflow applications with deadline constraints, and task scheduling. This chapter also presents a network energy-efficient workflow task scheduling algorithm that consists of task level computing, task subdeadline initialization, dynamic adjustment, and a data communication optimization method.

In Chapter 8, we present a novel resource-aware scheduling algorithm called RADS, which searches and deletes redundant task duplications dynamically in the process of scheduling. A further optimizing scheme is designed for the schedules generated by our algorithm, which can further reduce resource consumption without degrading the makespan.

Chapter 9 presents a novel contention-aware reliability management algorithm for parallel tasks in heterogeneous systems. Given that majority of previous studies do not consider the realistic existence of contention in modern communication systems, the algorithm is presented in the current study by applying DVFS and slack reclaiming techniques.

AUDIENCE AND READERSHIP

This book should be a useful reference for researchers, engineers, and practitioners interested in scheduling theory for computing systems. The book can be used as a supplement for graduate students and system developers whose major areas of interest are in resource management of cluster, supercomputers, grid computing, cloud computing, edge/fog computing systems, and related fields, as well as engineering professionals from both academia and computing systems development companies. By reading this book, readers will be familiar with new types of computing systems and their features, will learn a variety of scheduling algorithms, and find a source of inspiration for their own research.

ACKNOWLEDGMENTS

This book is published with financial support from National Key R&D Programs of China (Grant No. 2020YFB2104000), National Natural Science Foundation of China (Grant No. 61972146), Hunan Provincial Natural Science Foundation of China (Grant No. 2020JJ4376).

CHAPTER 1

Computing Systems

I N recent years, with the further promotion of computer technology and network technology, resource management and workflow scheduling are widely used in large-scale parallel and distributed computing systems, such as cluster systems, supercomputers, grid computing, cloud computing, edge/fog computing, and so on. However, there are many challenging issues in resource management systems and workflow task scheduling. This chapter first introduces the working principle of resource management and some typical resource management in computing systems. Then, this chapter presents practical applications of the DAG model and some real-world workflow applications.

1.1 COMPUTING SYSTEMS RESOURCE MANAGEMENT

With the rapid development of information technology, computing systems have become the core infrastructure of social economy. Resource management is one of the most common basic modules in such systems. However, as the requirement of computing amount and data storage increases sharply, resource management faces more challenging issues. In this book, resource management is the efficient and effective development of limited available components. It includes preventing resource leakage (when processes release resources) and processing resource contention (when processes access the same resource). In the current era of big data explosion, efficient resource management and task scheduling are essential methods to improve the performance of computing systems.

The main goal of resource management is to achieve efficient resource sharing. With the development and application of more and more efficient resource management systems, resource sharing efficiency of computing systems has been dramatically improved. Therefore, resource management and task scheduling have been paid great attention by industry and academia.

1.2 THE WELL KNOWN SYSTEMS

This section focuses on three typical resource management systems: Simple Linux Utility For Resource Management (SLURM), Portable Batch System (PBS), Yet Another Resource Negotiator (YARN).

1.2.1 SLURM

Simple Linux utility for resource management (SLURM) is an open source, fault-tolerant and highly scalable distributed resource management and job scheduling system, which is suitable for Linux clusters of all sizes [2, 122]. In addition, SLURM runs relatively independent and does not need to modify the kernel [129]. As a resource manager, SLURM has three key functions.

Firstly, SLURM allows users to have exclusive access to resources (computing nodes) for a period time to complete the required jobs. Secondly, it provides a framework to start, execute, and monitor job on a set of specified nodes (usually in parallel). Finally, it arbitrates resource contention by managing the queue of jobs waiting to be processed [2].

In addition, SLURM also has a plug-in mechanism, which allows developers to write plug-ins according to their own needs to implement SLURM extension. So far, there are some advanced functions in the newer version of SLURM, including fair sharing, backfilling, preemption, multi-priority and advanced reservation, etc. Today, SLURM has become the leading resource manager used on supercomputers, and some supercomputer centers even combine SLURM with other schedulers (such as Maui [19], LSF [164]) to achieve more efficient resource management. According to SLURM developers, about 40% of the TOP500 installations use SLURM.

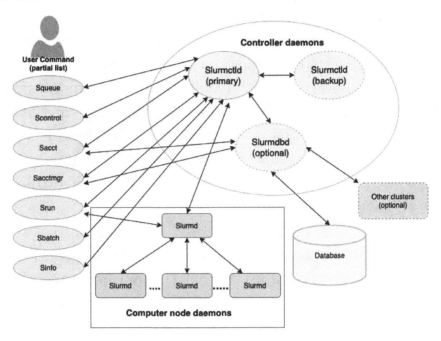

Figure 1.1 The main architecture of SLURM.

SLURM is the most popular resource management system at present, mainly because it has powerful functions and can solve unexpected situations in task scheduling. The main architecture of SLURM is shown in the Figure 1.1. SLURM has a centralized manager slurmctld to monitor resources. If a failure occurs, there may be a backup manager to complete it. Each computing server (node) has an slurmd daemon, which is like a remote shell: waiting for work, performing work, returning status, and waiting for the next work. What's more, the slurmd daemon provides fault-tolerant hierarchical communication. In addition, there is an optional slurmdbd (SLURM database daemon) that can be used to record the accounting information of multiple slurm-managed clusters in a single database. The slurmrestd (SLURM REST API daemon) can be used to interact with SLURM through its REST API. User tools include srun to start work, scancel to stop queued or running jobs, sinfo to report the status of partitions and nodes managed by SLURM, and squeue to report status jobs or job steps, sacct is used to summarize audit information for reporting activities or completed jobs and job steps. Sview is a graphical user interface for retrieving and updating the status information of jobs, partitions, and nodes managed by SLURM. There is a management tool scontrol that can be used to monitor and/or modify the configuration and status information on the cluster. The management tool used to manage the database is sacctmgr. Smap reports the status information of jobs, partitions, and nodes managed by SLURM, but displays the information graphically to reflect the network topology.

The entities managed by these SLURM daemons (as shown in Figure 1.2), include computing resource nodes (a group of logical computing resource combination partitions), resource allocation jobs assigned to users for a specified amount of time, and tasks in a group of jobs (maybe parallel tasks). The partitions can be regarded as job queues, and each queue has corresponding con-straints, such as user selection, job usage time limit, and job size limit. Then, the nodes in the queue are assigned to priority jobs until the queue's resources (such as nodes, processors, memory, etc.) are exhausted. After assigning a group of nodes to a job, the user can start the job as a job step in any configuration within the allocation. For example, a single job step can start using all nodes assigned to the job, or multiple job steps can independently use partial assignments. SLURM provides resource management for processors that have assigned jobs, allowing multiple job steps to be submitted and queued simultaneously until the resources in the job assignment are available.

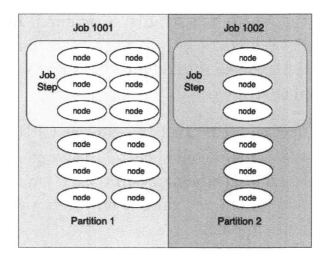

Figure 1.2 SLURM entities.

1.2.2 PBS

Portable Batch System (PBS) is jointly developed by the Numerical Aerospace Simulation (NAS) Systems Division of NASA's AMES Research Center and the National Energy Research Supercomputer Center (NERSC) at Lawrence Livermore National Laboratory [62]. PBS is a resource management and task scheduling software widely used in heterogeneous cluster systems. It has strong adaptability and high flexibility, and can provide a simple and consistent job submission method for job schedulers. It currently has two major versions: open-source OpenPBS [143] and commercial PBS Pro (PBS Professional) [6]. PBS Pro can distribute workloads across clusters of multiple platforms to simplify job submission, and its processors can be scaled to hundreds or even thousands.

Most modern systems require multiple processes to execute at the same time, and scheduling is a way to allow threads or processes to access system resources (e.g., processor time, which is usually used to balance the load on the system effectively [68]). It can be said that scheduling is the core of any workload management software. PBS Pro is a professional scheduling processing framework, which can select and schedule task execution according to the scheduling strategy configured by the user. In addition, it allows users to prioritize tasks to improve resource utilization and maximize throughput. PBS Pro supports the following scheduling strategies [12]

- First in, first out (FIFO). It maximizes CPU utilization. In FIFO, there are two coin sides. On the one hand, a thread can only run if it has the highest priority. On the other hand, a thread may be starved since it has been waiting for a long time without running.

- Job and queue loops. It is similar to FIFO but based on a configurable total amount of time. Within a cycle slice, its policy is based on priority.

- Fair sharing. It schedules jobs based on usage and shared values.

- Load balancing. It balances the load between time-sharing nodes and the nodes that pass through the loop.

- Dedicated time/node. It schedules jobs to specific nodes at specific times.

PBS Pro has four main components: Client commands (including submit, control, monitor, and delete jobs), which can be installed on any platforms that support qsub and qdel commands, as well as XPBS and other graphical tools [12]. Server (pbs_server) provides the main entry point for batch processing services, such as creating jobs, modifying jobs, and protecting jobs from system crashes. All clients and other daemons communicate with the server via TCP/IP. Scheduler (pbs_sched) controls the policies or rules used to submit jobs via the network. Initially, the scheduler queries the server for jobs to be run, and the query executor understands the availability of system resources. Moreover, each cluster can create its own scheduler or strategy. Job

Executor (pbs_mom) executes the job by imitating the user's session, and then sends the output to the caller. These four main components constitute the architecture of PBS (as shown in the Figure 1.3).

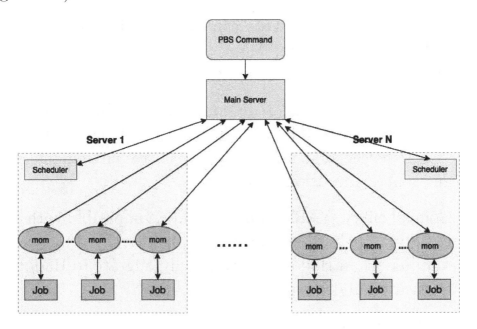

Figure 1.3 PBS structure.

1.2.3 YARN

Yet Another Resource Negotiator (YARN) is one of the core components of the open source Apache Hadoop distributed processing framework. It is mainly used for job scheduling and resource management of various applications in the cluster. YARN was originally called MapReduce 2, because it decouples the scheduling capabilities of resource management and data processing units in MapReduce by providing a new method, and promotes the original MapReduce to another level [37, 152]. In addition, YARN allows Hadoop developers to create applications that process large amounts of data, and use them in an effective way. In other words, YARN's architecture makes it more efficient and versatile than Hadoop MapReduce in big data processing, and also enhances the following aspects of Hadoop cluster [11].

- Multi-tenancy. YARN can access a variety of proprietary and open source engines. Therefore, interactive and batch processing tasks can be successfully deployed on Hadoop, and these tasks can access and parse the same data set.

- Cluster utilization. YARN can dynamically schedule the use of Hadoop clusters for MapReduce applications, which can make full use of the advantages of the cluster.

- Extensibility. YARN can provide scalability power to Hadoop clusters. The YARN ResourceManager (RM) service is the central controlling body for resource management and makes allocation decisions.

- Compatibility. YARN can be highly compatible with existing Hadoop MapReduce applications thereby projects using MapReduce in Hadoop 1.0 can easily migrate to Hadoop 2.0 to ensure full compatibility.

YARN is designed to split JobTracker in MRV1 into two independent services: Global ResourceManager and ApplicationMaster specific to each application. ResourceManager is responsible for the management and allocation of overall system resources, and ApplicationMaster is responsible for the management of each application [87]. In fact, YARN is still a master/slave structure as a whole, with a ResourceManager as the Master and a NodeManager as the Slave (Figure 1.4 shows the overall architecture of YARN).

ResourceManager (RM) is the process on the master that is responsible for resource management and scheduling for the entire distributed system. It processes requests from clients (including submitting jobs/killing jobs); starts/monitors ApplicationMaster; monitors NodeManager conditions, such as NodeManager that may be down [64].

NodeManager (NM) is the process on the slave node, which is only responsible for the resource management and scheduling of the current slave node, as well as the operation of the task. It periodically reports resource/container status (heartbeat) to ResourceManager, and receives start and stop commands from ResourceManager Container [64].

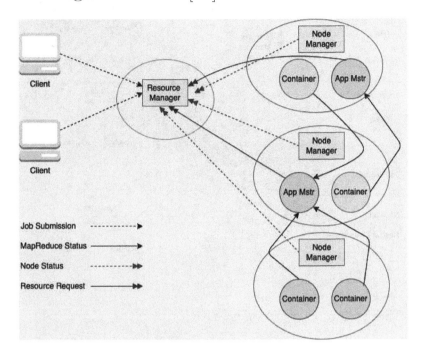

Figure 1.4 The structure of YARN.

Each job submitted to the cluster will have a corresponding ApplicationMaster (AM) responsible for managing applications and data segments; requesting resources from the ResourceManager (also known as Containers) of the current application and assign them to specific tasks. ApplicationMaster communicates with NodeManager to start and stop specific tasks running in the Container. Task monitoring and fault tolerance are also the responsibility of ApplicationMaster [64].

After recognizing the major components of YARN, the implementation of Application in YARN can be summarized in three steps: First, application submission. Second, start the ApplicationMaster instance of the application. Finally, ApplicationMaster instance manages the execution of the application. Figure 1.5

shows the entire execution of the application, which includes the following steps

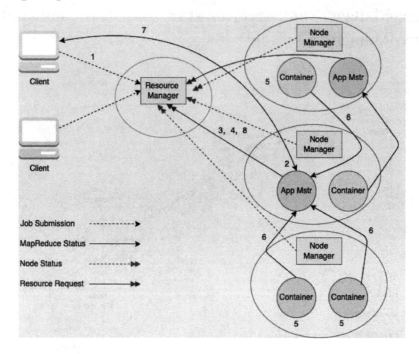

Figure 1.5 The application execution on YARN.

1. The client application submits an application to the ResourceManager and requests an ApplicationMaster instance.

2. The ResourceManager finds a NodeManager that can run a Container and starts an ApplicationMaster instance in this Container.

3. The ApplicationMaster registers with the ResourceManager, and after registration the client can query the ResourceManager to get the details of its ApplicationMaster, and later it can interact directly with its own ApplicationMaster.

4. In the ordinary course of operation, the *ApplicationMaster* sends resource-request requests to the ResourceManager according to the resource-request protocol.

5. After the Container is successfully allocated, the ApplicationMaster starts the Container by sending container-launch-specification information to the NodeManager, container-launch-specification The container-launch-specification message contains

the information needed to enable the Container to communicate with the ApplicationMaster.

6. The application code runs in the launched Container and sends the running progress, status and other information to the ApplicationMaster via the application-specific protocol.

7. During the running of the application, the client submitting the application takes the initiative to communicate with the ApplicationMaster to get the running status, progress updates and other information of the application, and the protocol of communication is also application-specific protocol.

8. Once the application execution is completed and all related work has been done, the ApplicationMaster unregisters with the ResourceManager and then shuts down, and the Containers used will be returned to the system.

YARN is designed to allow our various applications to use the entire cluster in a shared, secure, and multi-tenant manner, and it must also be able to perceive the entire cluster topology to ensure efficient cluster resource scheduling and data access. In order to achieve these goals, ResourceManager scheduler defines a flexible protocol for application resource requests to better schedule various applications running in the cluster, resulting in *Resource Request* and *Container* [152]. For more information, please refer to relevant books.

1.3 PARALLEL APPLICATIONS

Nowadays, large-scale high-performance or transaction processing parallel applications usually have thousands of tasks with mutual priority constraints and computational dependencies. These applications require tens or millions of processing cores to compute. The popular representation of these parallel applications is a directed acyclic graph (DAG) in which the nodes represent application tasks and the directed arcs or edges denote inter-task dependencies, such as task's precedence constraint. We also call DAG as workflow application.

1.3.1 Workflow Applications

In computing systems, resources (such as infrastructure, platforms, and software) are allocated to computing applications, and these computing applications (e.g., scientific workflows) often require large amounts of resources from various computing infrastructures to process large amounts of big data in systems. With the rapid development of cloud computing, edge computing, high-performance computing, applications are gradually evolving toward these systems. The tasks of these applications are usually precedence constraints and we referred to as workflow. Many workflows are usually modeled as a set of tasks that are connected to each other through data or computational correlations. Tasks in a workflow are linked according to their computational dependencies, and these tasks are represented as directed acyclic graphs (DAGs), and workflows with priority constrained tasks are increasing with the continuous expansion of heterogeneous systems [75, 146]. These workflow applications are often used to solve large-scale scientific problems in a various fields such as bioinformatics, astronomy, and physics.

1.3.2 Classical Tasks DAG Model

Task graphs are encountered in many models used in operations research and resource management, such as scheduling, computer and network models. Graph models are often used to study the activities that contain some concurrency and precedence constraints. For example, the DAG can be described as workflow parallel applications, in which nodes represent tasks and directed arcs represent synchronization constraints, data communication among tasks.

Generally, a workflow application is represented by a DAG model $G = \langle T, A \rangle$, where T is the set of $|T|$ tasks that can be scheduled to execute on any of the available machines, cores, processors, cloud virtual machines (VMs), cloud containers, and so on. $A \in T \times T$ is the set of directed edges between the tasks to represent the priority dependencies [146]. For instance, DAG edge $a_{i,j} \in A$ represents the priority constraint such that task t_i should finish its execution before the task t_j starts its execution. A task may have one or more priority constraints. In this model, the task can be triggered to execute when all prioritization tasks and corresponding computational resources are available.

The weight $w(t_i)$ assigned to task t_i represents its execution size (such as millions of instructions, MIs), and the weight $w(a_{i,j})$ assigned to DAG edge $a_{i,j}$ represents its data communication requirements. Numerous studies have addressed the topic of estimating task characteristics, such as analytical benchmarking, historical table, code profiling, *SKOPE*, and statistical probabilistic techniques [140]. The *SKOPE* is an effective technique that the user can submit parallel application code skeleton including data flow, control flow, functions, data set, and computational intensity. According to the semantics and the structure in the code skeleton, the *SKOPE* back-end can explore task execution size and data communication between tasks [140].

The set $\{t_j \in T, a_{j,i} \in A\}$ of all direct predecessors of t_i is denoted by prev(t_i) and the set $\{t_j \in T, a_{i,j} \in A\}$ of all direct successors of v_i is expressed as next(t_i). Normally, a DAG node or task $t \in T$ without predecessors, which we call it as *entry* task (t_{entry}). On the other hand, if the succ(t_i) is null, this book also names it as *exit* task (t_{exit}). Without losing of generality, in this book, it assumes that one workflow application DAG has exactly one t_{entry} and one t_{exit}. If multiple entry tasks or exit tasks exist, they may be connected with zero weight edges to a single pseudo-t_{entry} or a single t_{exit} that has zero weight. Figure 1.6 shows an example DAG with assigned tasks and precedence constraint edges.

1.4 SOME REAL-WORLD WORKFLOW APPLICATIONS

The scientific community is using workflow technology to manage complex and intensive data simulations and analysis. Workflow

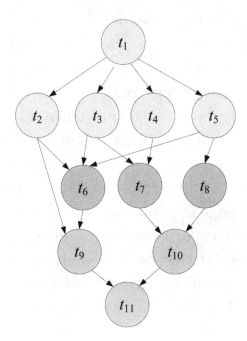

Figure 1.6 An example of workflow application DAG model.

technologies are responsible for managing the dependencies between tasks and scheduling computational tasks based on these
dependencies [146,157]. Many workflows are typically modeled as
a set of tasks that are interconnected by data or computational
dependencies. The workflow parallel applications are often used
to solve large-scale scientific problems in a variety of fields, such
as bioinformatics, astronomy, and physics [69,111,112]. Several
representative scientific workflows are briefly described as follows.

1.4.1 Montage

Montage is essentially a set of derivatives or channels created by
NASA/IPAC Infrared Science Archive, which can generate cus-
tom mosaics of the sky by using input images in the Flexible
Image Transmission System (FITS) format [69,112]. During the
generation process, the input image determines the geometry of
the output image, and then the input image is reprojected and ro-
tated with the same spatial scale and rotation. If the background
emission in the image is corrected to a uniform level, then the

re-projected, corrected image will be added to form a mosaic. In addition, most tasks of montage are characterized by being I/O intensive and do not require too much CPU processing capacity. Therefore, montage is a typical representative of I/O-bound workflows [112].

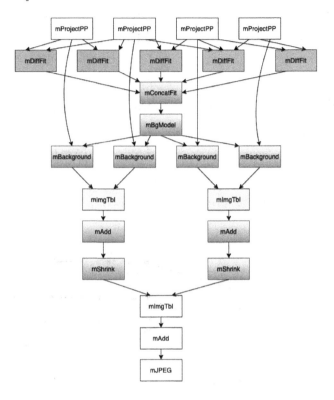

Figure 1.7 Montage workflow.

A classic montage workflow DAG is shown in Figure 1.7. At the top of the workflow shown is the mProjectPP job, which re-shapes the input image. The mBgModel and mConcatFit jobs are data segmentation and data aggregation jobs, respectively. The number of mProjectPP jobs is equal to the number of FITS input images processed. The reprojected image and the area image are the output of these jobs. For the next pass, mDiffFit is used to calculate the difference between each pair of overlapping images. Then, the mConcatFit job fits the difference image. Next, in order to obtain a good global fit, the mBgModel job calculates the correction for each image. In the next process, the mBackground job applies the correction to each image. The mImgTbl job aggregates the metadata of all images. The reprojected image is recombined by the mAdd job to generate mosaic and area images in FITS format. Multiple levels of mImgTbl and mAdd jobs can be used in large workflows. Finally, the mShrink job modifies the size of the FITS image and converts the modified image from mJPEG to the JPEG format [69].

1.4.2 Broadband

Broadband is the computing platform used by the Southern California Earthquake Center [18]. Broadband's goal is to integrate computational integration and motion simulation code to generate discoveries that are valuable to seismic engineers. These codes form a workflow that simulates the effects of earthquakes on records.

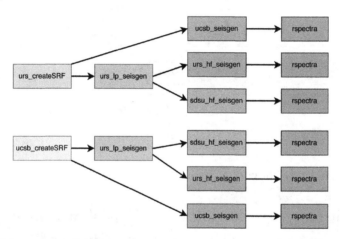

Figure 1.8 Broadband workflow.

Figure 1.8 is an example broadband workflow DAG. The first stage of rupture generation operations ucsb_createSRF and urs_createSRF operations both use the same earthquake description as input to form the same time series data station that records the time history of earthquake jumps. The second stage of the urs_lp_seisgen operation calculates deterministic low-frequency (up to 1 Hz) seismograms. The third stage includes sdsu_hf_seisgen work, adding random high-frequency seismograms to the low-frequency earthquakes generated in the second stage ucsb_seisgen work mixed with low frequencies and high-frequency seismograms can be considered as part of the second and third phases. The fourth stage package rspectra job to extract the parameters of interest to the seismic engineer from the seismogram [18].

1.4.3 Epigenomics

The epigenomics workflow is essentially a data processing pipeline that uses the Pegasus workflow management system to automate various genome sequencing operations [69, 111, 112]. The DNA sequence data generated by the Illumina Solexa Genetic Analyzer System is divided into blocks that can be manipulated in parallel. The data in each block is converted into a file format that can be used by the Maq software to map short DNA sequencing reads. From there, the sequences are filtered to remove noise and contaminating fragments and are mapped to the correct position in the reference genome. Finally, a global map of the compared sequences is generated and the sequence density is calculated for each position in the genome [69]. Contrary to montage, Epigenomics workflow is a CPU-intensive application [112].

An example of epigenomics workflow is shown in Figure 1.9. The input to this workflow is DNA sequence data obtained for multiple lanes in the genetic analysis process. At beginning, the information from each lane is split into multiple chunks by the fastQSplit job. Then the filterContams job processes away the noisy and contaminated data from each block. After that, the data is then converted to binary fastQ format by fastq2bfq. Next, the remaining sequences are compared to the reference genome by the map tool. The results of the individual map processes are combined using one or more stages of the mapMerge job and the mapIndex tool operates on the merged alignment file. At last, the pileup tool reformats the data and displays it by the GUI [69].

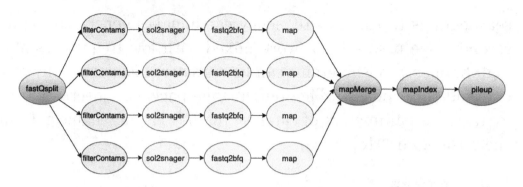

Figure 1.9 Epigenomics workflow.

1.4.4 LIGO Inspiral Analysis

The Laser Interferometer Gravitational-Wave Observatory (LIGO) is attempting to detect gravitational waves generated by various events in the Universe [69, 112]. The LIGO Inspiral Analysis workflow is used to analyze data obtained from the coalescence of dense binary systems. Time-frequency data from the three LIGO detectors are divided into smaller chunks for analysis [69]. For each block, the workflow generates a subset of waveforms belonging to the parameter space and calculates the matched filter output. If a real detector is detected, a trigger is generated that can be checked with triggers from other detectors. Several additional consistency tests can also be added to the workflow [69].

Figure 1.10 shows an approximate structure of a typical gravitational physics scientific application LIGO [69]. The LIGO

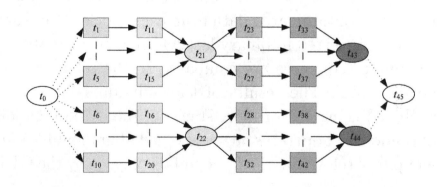

Figure 1.10 LIGO workflow.

Inspiral Analysis application is very complex and mainly composed of Data Find, Inspiral, TmpltBank,Thinca, Inca, Inspinj, TrigBank, Sire, and Coire sub-workflows. Most of LIGOs sub-workflows are computationally intensive tasks. For example, the Inspiral CPU utilization achieves 89.96% [69, 112].

1.5 OUTLINE OF THE BOOK

The outline of this book is below.

- Chapter 2 first introduces the scheduling problems, workflow task scheduling, scheduling challenges, and the classification of scheduling algorithms. Then, we present several typical heuristic scheduling algorithms.

- In Chapter 3, we drive a stochastic scheduling model on grid computing systems. In order to effectively scheduling precedence constrained stochastic tasks, this chapter present a stochastic heterogeneous earliest finish time scheduling algorithm, which incorporate the stochastic attribute, such as expected value and variance, of task processing time and edge communication time into scheduling.

- Chapter 4 emphasizes the scheduling stochastic parallel applications with precedence constrained tasks on heterogeneous cluster systems. It presents a stochastic dynamic level scheduling algorithm, which employs stochastic bottom level and stochastic dynamic level to produce schedules of high quality.

- Chapter 5 first builds a reliability and energy aware task scheduling architecture. Then, this chapter constructs a single processor failure rate model based on DVFS technique and deduce the application reliability of systems. Finally, we present a heuristic reliability-energy aware scheduling algorithm.

- In Chapter 6, we address the bi-objective optimization problem of high system reliability and low energy consumption for parallel tasks as a combinatorial optimization problem.

- Chapter 7 addresses the issues of heterogeneous systems, energy consumption of processors and interconnection networks, computation-intensive scientific workflow applications with deadline constraints, and task scheduling. This chapter presents a network energy-efficient workflow task scheduling algorithm.

- In Chapter 8, we present a resource-aware scheduling algorithm, which searches and deletes redundant task duplications dynamically in the process of scheduling.

- Chapter 9 describes a contention-aware reliability management algorithm for parallel tasks in heterogeneous systems.

CHAPTER 2

Classical Workflow Scheduling

THIS chapter introduces task scheduling, scheduling challenges, and the classification of scheduling algorithms. It also presents several typical heuristic scheduling algorithms, such as Dynamic Level Scheduling (DLS), Modified Critical Path (MCP), and Earliest-Finish-Time (HEFT). This chapter hopes that the readers have a preliminary understanding of the workflow scheduling concepts of computing systems resource management.

2.1 TASK SCHEDULING

The task scheduling problem is essentially a combinatorial optimization problem. The optimal solution of the combinatorial optimization problem is an NP (non-deterministic polynomial)-hard problem, especially multi-objective task scheduling is an NP-hard problem that satisfies the security of the system, and the reliability between tasks and constraints [103].

The scheduling system has two fundamental evaluation indicators: performance and efficiency. They are based on the quality of task allocation (scheduling) and the efficiency of the scheduling algorithm (the scheduler itself). The scheduling quality is measured by the performance of the generated optimal scheduling,

DOI: 10.1201/b23006-2

and the scheduling efficiency is measured by the time complexity of algorithm itself. For example, scheduling quality is measured by the completion time of the optimization process [103]. Obviously, the shorter the completion time, the better the scheduling algorithm. If two scheduling algorithms produce the same scheduling quality, the simpler one is better.

2.2 SCHEDULING CHALLENGES

In recent years, computing systems are facing many challenging issues, such as energy consumption, reliability, real-time, and so on. To deal with these problems, the corresponding scheduling strategies have been presented that can be broadly classified into the following categories.

2.2.1 Energy-Efficient Scheduling

Energy management is essential in modern parallel and distributed systems because various adaptive management techniques are required to maximize energy efficiency and solve thermal problems. However, different computing and data storage levels require high energy consumption, which is a major problem affecting the development of computing systems and human environmental protection. Solving this problem requires more human resources, material resources, and financial resources. DVFS is one of the methods to achieve energy-saving scheduling. It saves energy consumption by simultaneously reducing the power supply voltage and frequency of the processor, thereby reducing environmental damage [78].

2.2.2 Reliability-Aware Scheduling

Due to various configurations or capacities of hardware or software in computing systems, reliability is a considerable challenge. With the increase in the scale of computing, the diversity of equipment has also become complicated. Thereby the incidence of transient failures increases dramatically when workflows are executed on such systems [26]. Therefore, reliability-aware scheduling was used to deal with this type of issue.

Reliability is defined as the probability of successfully completing the execution of the plan. It has been widely regarded as an increasingly important issue in computing systems [106, 177]. In practice, many applications cannot meet their reliability requirements 100%, so several reliability-related standards have appeared worldwide. These can also as the QoS standard of computing system [175]. If an application can meet its reliability requirements, then it is considered reliable. In addition, the fault tolerance of master and backup replication is an important reliability enhancement mechanism. In the master-backup replication scheme, a master task will have one or more backup tasks, and other backups are collectively referred to as copies [17, 176]. Therefore, it is necessary to meet the reliability requirements from the perspective of standards and service quality.

2.2.3 High Performance Real-Time Scheduling

Traditional real-time scheduling usually considers scheduling problems from the task level. With the application of new models and more classical scheduling theories to relatively new platforms and applications, users have higher requirements for high performance and real-time performance. However, these two requirements are in conflict with each other in real-time scheduling for multiple applications.

From the point of view of high performance, minimizing the total scheduling length of the system is the main requirement [63]. Meeting application deadlines is one of the most important security requirements in terms of time constraints. In general, fair strategies aim to reduce the total scheduling length of the system or the individual scheduling length of the application, but in this case the system cannot meet the requirements of all applications.

2.3 SCHEDULING ALGORITHMS CLASSIFICATION

The types of task scheduling algorithms are defined differently according to different classification methods. Among the algorithms proposed for scheduling tasks, the following main categories can be identified: local and global, static and dynamic, online and offline, optimal and suboptimal, approximate and heuristic, preemptive and non-preemptive, centralized and distributed, cooperative and non-cooperative, etc. Various scheduling strategies are shown in Figure 2.1 and Table 2.1.

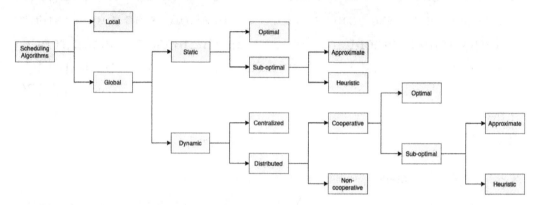

Figure 2.1   A hierarchical taxonomy for task scheduling.

2.3.1   Local versus Global

In level scheduling, local scheduling rules determine how to allocate and execute processes running on a single CPU. The global scheduling optimizes the performance of the system by using the relevant information of the system to allocate processes to the waiting processors.

2.3.2   Static versus Dynamic

Static algorithms are those algorithms whose scheduling decisions are based on fixed parameters and are assigned to tasks before they are activated. Dynamic algorithm refers to an algorithm based on dynamic parameters that may be changed in the process of system evolution.

TABLE 2.1 Taxonomy of task scheduling strategies

Categories	Division reasons	Typical algorithms	Characteristics
Local	Level scheduling	Proportional Sharing [132], Predictive [115]	Local scheduling is involved with the assignment of processes to the time-slices of a single processor.
Global	Level scheduling	SJF [99], HEFT [117]	Global scheduling is the problem of deciding where to execute a process.
Static	Allocation of resource	FIFO [80], SJF [99]	Static scheduling performs better when the workload is not varying frequently and the variation in system behavior is very little
Dynamic	Allocation of resource	HEFT [117], WLC [109]	Dynamic scheduling needs no advance information about the node (VM) and task; however, the node requires continuous monitoring.
Optimal	Integrity	EDF [89], LLF [38]	Optimal scheduling is the optimal allocation according to some criterion functions under the condition that resource state and job state information are known.
Suboptimal	Integrity	Log Scheduling [39], Wave Scheduling [144]	Suboptimal scheduling is due to the fact that it is difficult to make reasonable assumptions in the actual situation, and often attempts to find suboptimal solutions.
Approximate	Reasonable	Polynomial Time Algorithm [88]	Approximate scheduling uses formal computational models, it is satisfied when a sufficiently good solution is found.
Heuristic	Reasonable	SA [73], ACO [45]	Heuristic scheduling is a kind of scheduling that makes the most realistic assumptions about a priori knowledge concerning process and system loading characteristics.
Centralized	Priority basis	C-PSGD [5]	Centralized scheduling means that all tasks are collected by the main processor unit and each processor takes over a separate scheduling queue.
Distributed	Priority basis	Load Balancing Algorithms [91]	Distributed scheduling refers to the fact that the local scheduler continuously shares information with other schedulers to update and maintain their scheduling status.

2.3.3 Optimal versus Suboptimal

If the resource usage and job information are known, the allocation based on certain standard functions (such as the fastest completion time and the maximization of resource utilization) is called optimal. However, due to the incompleteness of the scheduling algorithm and the difficulty of making correct assumptions to prove the superiority of the algorithm in real scenarios, researchers try to find the suboptimal of the algorithm, which can be further divided into the following two categories.

2.3.4 Approximate versus Heuristic

The approximate algorithms use formal computational models, but instead of searching the entire solution space for an optimal solution, they are satisfied when finding enough good solutions. Heuristics algorithms represent the category of algorithms that make the most realistic assumptions about the prior knowledge of the process and system load characteristics. It also represents a solution to the scheduling problem that cannot give the best answer but only needs the most reasonable cost and other system resources to perform its functions.

2.3.5 Centralized versus Distributed

In centralized scheduling, each task is collected in a master processor unit and then sent to the slave processor unit, where each processor also has an independent scheduling queue. In distributed scheduling, the local server is responsible for all tasks, including processing incoming requests and maintaining the state of all processors by sharing information with other processors.

2.4 SEVERAL HEURISTIC WORKFLOW SCHEDULING ALGORITHMS

In many computing systems, due to the diversity of computing resources, the efficiency of an application over the available resources is one of the key factors in achieving high performance computing. The primary goal of task scheduling is to assign tasks to appropriate resources and to rank tasks to meet the task priority with minimum scheduling length [146].

Typical task scheduling algorithms are based on DAG model including static scheduling and dynamic scheduling [60, 182]. In static scheduling algorithms, all the information needed for scheduling, such as the structure of the parallel application, the execution time of individual tasks [158], etc., must be known in advance. Static task scheduling occurs during the compilation period before running the parallel application. In contrast, in dynamic scheduling algorithms, tasks are assigned to the processor on arrival and scheduling decisions are made at runtime [60, 75]. Heuristic scheduling algorithms are static scheduling algorithms, which can be divided into three main categories: repetition heuristics, list-based heuristics, and clustering heuristics [146].

There are many effective heuristics scheduling algorithms such as Mapping Heuristic (MH) [49], Dynamic Critical Path (DCP) [74], Levelized Min Time (LMT) algorithm [66], Modified Critical Path (MCP) [158], Dynamic Level Scheduling (DLS) algorithms [120], Critical Path on a Processor (CPOP) algorithms, and Heterogeneous Earliest-Finish-Time (HEFT) algorithm [146]. There are also heuristic algorithms that address the optimality of task scheduling in special cases, which are models with special constraints, such as Tree models [167] and some other task graph models that do not account for overhead, etc. This section focuses on three classical heuristic workflow scheduling algorithms based on the DAG task model: DLS, MCP, HEFT.

2.4.1 DLS

The DLS algorithm is an efficient compile-time scheduling technique that takes into account the interprocessor communication overhead when mapping priority-constrained communication tasks onto an arbitrarily interconnected network of processors

[120]. What's more, it can also schedule acyclic priority graphs to multiple processor architectures with limited or irregular interconnect structures and reduce execution time. When making scheduling decisions, the DLS algorithm takes into account the heterogeneity of resource nodes, so as to effectively adapt to different resources in computing systems. For example, an improved DLS algorithm, called R-DLS algorithm, is proposed in the paper considering the risk-taking level of resource nodes [160].

The DLS algorithm uses a dynamic level DL, which is the difference between the static level of a node and its earliest start time on a processor. At each step of the scheduling, the algorithm computes the dynamic level DL (t_i, p_j) for each node in the ready pool on all processors. t denotes the number of nodes and p represents the number of processors [120]. The dynamic level of a task-machine (t_i, p_j) is defined to be

$$DL(t_i, p_j) = SBL(t_i) - max\{t_{i,j}^A, t_j^p\} + \Delta(t_i, p_j), \quad (2.1)$$

where $SBL(t_i)$ represents the static level of the task, $max\{t_{i,j}^A, t_j^p\}$ is the time when task t_i can begin execution on machine p_j, $t_{i,j}^A$ denotes the time when the data will be available if task t_i is scheduled on machine p_j, t_j^p denotes the time when machine p_j will be available for the execution of task $t_i \cdot \Delta(t_i, p_j) = t_i^E t_{i,j}^E$ reflects the computing performance of the machine, t_i^E shows the execution time of the task t_i on all the free machines, and $t_{i,j}^E$ represents the execution time of task t_i on machine p_j.

Before understanding the algorithm, this section presents several definitions. In a DAG, $t\_level(t_i)$ refers to the longest path length from the entry node to the node t_i. The path length here refers to the sum of the weights of all nodes and edges on this path. Including the weight of t_i itself; $b\_level(t_i)$ refers to the longest path length from node t_i to the exit node, and the b_level of a node will not exceed the length of the DAG critical path.

ALAP (as-soon-as-possible) refers to how long a node's start time can be postponed without affecting task scheduling. SBL refers to the calculation of *b_level* without considering the weight of the edge, so that *b_level* is a constant throughout the task scheduling process, so it is called static *b_level* (SBL).

DLS selects the (node, processor) pair with the largest dynamic level for scheduling. This mechanism is similar to the mechanism used by the Earliest Time First (ETF) algorithm [75]. The subtle difference is that the ETF algorithm always schedules the node with the earliest start time and adjusts it for the static level b. On the other hand, the DLS algorithm tends to schedule nodes in descending order of dynamic level at the beginning of the scheduling process, and schedule nodes in ascending order of level (i.e., the earliest start time) near the completion of the scheduling process. DLS needs to provide a routing table to complete the dispatch of messages. The steps of the DLS algorithm are as follows:

- Calculate the SBL value for each node.

- Initialize the pool of ready nodes to contain all the entry nodes; then iterate the following steps until all tasks have been scheduled.

- Calculate the earliest start time of each node on each processor, and then subtract the node's SBL from the earliest start time. The DL of each node and processor pair is calculated by subtracting the node's SBL from the earliest start time.

- Select the node, processor pair with the largest DL and schedule that node to the corresponding processor.

- Add the nearest ready node to the pool of ready nodes.

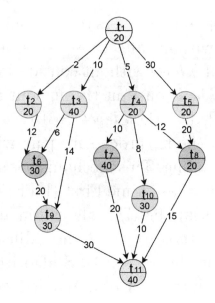

Figure 2.2 DAG task diagram example.

The time complexity of the DLS algorithm is $O(p \times v^3)$, where p denotes the number of processors, v signifies the number of task nodes, and e represents the number of edges in the DAG. For the DAG task graph shown in Figure 2.2, the Table 2.2 lists several attribute values of each node. Figure 2.3 shows a simple arbitrary processor network topology. The task scheduling sequence generated by the DLS algorithm in its system application is: t_1, t_3, t_2, t_4, t_6, t_5, t_7, t_{10}, t_9, t_8, t_{11}, its makespan is 47.33.

2.4.2 MCP

Similar to the DLS, the Modified Critical Path (MCP) algorithm first calculates the latest possible scheduling time ALAP of all nodes [158]. Then, it generates a list of each node including its children, which decreases according to ALAP. After constructing a task list according to the increasing order of ALAP, the nodes in the task list are sequentially scheduled to the necessary machine with the earliest scheduling time by using the insertion method. The steps of the MCP algorithm are as follows:

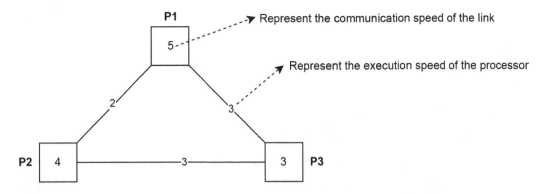

Figure 2.3 A simple arbitrary processor network topology diagram.

TABLE 2.2 The attribute value in Figure 2.2

Node	SBL	*t_level*	*b_level*	ALAP
t_1	160	0	226	0
t_2	120	22	182	44
t_3	140	30	196	30
t_4	100	25	130	96
t_5	80	50	115	111
t_6	100	76	150	76
t_7	80	55	100	126
t_8	60	90	75	151
t_9	70	126	100	126
t_{10}	70	53	80	146
t_{11}	40	186	40	186

- Perform the ALAP binding and assign the resulting ALAP time $T_L(t_i)(i = 1, 2, ..., n)$ to each node in the graph.

- Create a list l to sort each node t, and then create a node list L in descending lexicographical order.

- Schedule the first node in L to the processing element that allows it to execute earliest. Remove the node from L and repeat the last step until the L list is empty.

The time complexity of the MCP algorithm is $O(n^2 \log n)$, n is the number of task nodes. Combining the ALAP in the Table 2.2, the scheduling sequence generated by the MCP algorithm on the Figure 2.3 is: t_1, t_3, t_2, t_4, t_6, t_7, t_5, t_9, t_{10}, t_8, t_{11}, and its makespan is 47.33.

2.4.3 HEFT

The HEFT algorithm is a basic static scheduling algorithm for a limited number of heterogeneous processors for application scheduling [146]. Its purpose is to schedule all tasks to be executed on the processor that will enable it to finish earliest. The procedure is to calculate the earliest completion time EFT of all tasks on each processor before scheduling according to the dependencies in the DAG, and then compare the minimum earliest completion time and the processor that can satisfy this condition.

The HEFT algorithm has two main phases: task priority phase and processor selection phase [146]. In the task priority ranking stage, the priority of each task in this stage is set according to the upward ranking value of the average calculation and the average communication cost, and the task list is generated by sorting the tasks from high to low according to the ranking value. The descending order of rank values provides the topological order of tasks, which is linear to maintain priority constraints. The randomness of the tie process can effectively break the tie as an alternative strategy. The method is

$$rank(t_i) = \overline{w}_i + max_{t_j \in succ(t_i)}(\overline{c}_{i,j} + rank(t_j)), \qquad (2.2)$$

$$rank(t_{exit}) = \overline{w}_{exit}, \qquad (2.3)$$

$$\overline{c}_{i,j} = \overline{L} + data_{i,j}/\overline{B}, \qquad (2.4)$$

where $\overline{w}_i$ represents the average computational cost of task t_i, $succ(t_i)$ is the set of direct successor nodes of task t_i, and $\overline{c}_{i,j}$ is edge $e_{(i,j)}$ average communication overhead. $\overline{L}$ denotes the average communication startup time, $data_{i,j}$ represents the amount of data transferred from task t_i to task t_j, and $\overline{B}$ signifies the average transfer rate between processors.

In the processor selection phase, the HEFT algorithm has an insertion-based strategy that considers the possibility of inserting a task in the earliest idle time slot between two already scheduled tasks. The length of the idle time slot, i.e., the difference between the execution start time and completion time of two tasks scheduled consecutively on the same processor. And, the scheduling on this idle time slot should retain the priority constraint. Thus the earliest available time for a processor to execute a task is not the time when the processor finishes executing its last assigned task.

The HEFT algorithm steps are as follows:

- Calculate the average computing cost and average communication cost of each task node.

- Use the above formula to recursively calculate the priority of each task node.

- Repeat the following steps until all task scheduling ends. (1) Select the first task t_i from the task priority list. (2) Calculate the earliest completion time of tasks scheduled to other processors by using an insertion-based scheduling strategy. (3) Schedule the task to the processor with the earliest completion time.

The time complexity of HEFT is $O(m \times v^2)$, where m is the number of processors and v is the number of tasks. For the DAG task graph shown in Figure 2.3, the scheduling order generated by the HEFT algorithm in its system application is: t_1, t_3, t_2, t_6, t_4, t_5, t_9, t_7, t_{10}, t_8, t_{11}, and its makespan is 43.67. HEFT algorithm significantly outperforms DLS, DCP, LMT, MH, MCP, and CPOP algorithms in terms of average makespan, speedup, etc. [146].

2.5 SUMMARY

This chapter presents the task scheduling, scheduling challenges, taxonomy of scheduling algorithms, and some heuristic workflow scheduling strategies.

CHAPTER 3

Stochastic Task Scheduling on Grid Computing Systems

THIS chapter discusses how to schedule the workflow application with precedence constrained tasks, random tasks execution size, and random data communication, on grid systems, which the goal is to minimize its schedule length. This is an interesting and difficult problem on distributed, dynamically, and geographically wide area computing systems. We formulates them in a form of stochastic scheduling model. A stochastic heterogeneous earliest finish time (SHEFT) scheduling algorithm is also presented, which incorporates the variance and expected value of stochastic processing time into the scheduling.

3.1 INTRODUCTION

At present, with the popularization of the Internet, high-speed interconnection networks, availability of powerful computers as low-cost commodity components, it becomes a reality to build large-scale grid computing systems. Because of these technologies, heterogeneous resources distributed in different regions can realize selection, aggregation, and sharing to solve large-scale problems in commerce, science, and engineering etc. [53, 156]. In order to achieve the promising potentials of enormous distributed resources, it is fundamentally important to build effective and

DOI: 10.1201/b23006-3

efficient scheduling algorithms. The scheduling problem deals with the allocation and coordination of computing resources so as to execute the parallel applications [136].

In grid systems, applications are submitted by users and generally independent each other, which request systems computing resources for their execution. Workflow application is consisted of tasks that usually require the use of different types of resources, e.g., storage, computation, communication, or specific instruments [136]. Scheduling aims at meeting user demands (e.g., in terms of schedule length, maximum lateness, sum of weighted completion times) and the goals denoted by the resource providers (e.g., in terms of resource utilization efficiency, profit), while maintaining a good overall performance for the grid [27]. As we all know, the optimal scheduling problem is NP-hard in the general case [55]. In general, we prefer to use heuristic methods to obtain scheduling solution rather than most costly optimal scheduling. These methods may obtain suboptimal results, but they are much computational cheaper [136].

Heuristic list scheduling is a very popular scheme for precedence constrained workflow DAG tasks scheduling. Its basic idea is to assign priority to tasks in DAG and arrange tasks in a list according to the order of priority [136]. Obviously, high-priority tasks will be scheduled first, and some way will be used to break the association between tasks. An important issue in workflow application DAG tasks scheduling is how to rank the tasks and edges (when communication delay is considered) [136]. After the DAG model tasks and edges are ranked, task-to-resource allocation minimize the schedule length by considering solving two problems: how to parallelize those tasks having no precedence orders in the DAG application graph, and how to make the time cost along with the DAG model critical path as small as possible [136].

Most of the above studies assume that the parameters such as task execution size, and data communication between tasks are fixed and deterministic, and these parameters are known in advance [136]. However, in practical problems, it is usually not enough to find a good schedule solution for fixed deterministic task execution size, because workflow application tasks usually contain some conditional instructions, which may have different processing time for different inputs [97,105,119,145]. The method to solve this problem is to use stochastic scheduling. In stochastic scheduling, the task processing time and data communication between tasks are interpreted as random variables, and then the performance of the algorithm is measured by their expected objective-value [136].

In this chapter, we let $T = \{t_1, \ldots, t_n\}$ be a set of tasks in a parallel workflow DAG that have to be scheduled on grid systems with m processors so as to minimize the makespan [136]. Any processor can execute at most one task at a time, and every task has to be processed on one of the m processors. In contrast to the deterministic task scheduling problem, the key assumption is that the task execution size $w(t_i)$ and data communication $w(a_{i,j})$ are not known in advance. Here, we also assume that the task execution size and data communication requirements are random variables, from which we can get their distribution [136].

3.2   THE GRID SCHEDULING ARCHITECTURE

Grid computing systems are highly diverse systems, and geographically distributed nodes are interconnected through wide area network (WAN). The grid node contains many computing resources with different processing capabilities. Heterogeneity and dynamics cause resources in grids to be clusters or distributed, rather than uniform distribution [136]. This chapter summarizes the resource management and job scheduling process in grids into two stages: resource discovery and filtering, resource selection, and scheduling jobs according to specific objectives. This chapter focuses on the second step: scheduling jobs. Here, we build a architecture of grid scheduling systems, which is shown in Figure 3.1 [136].

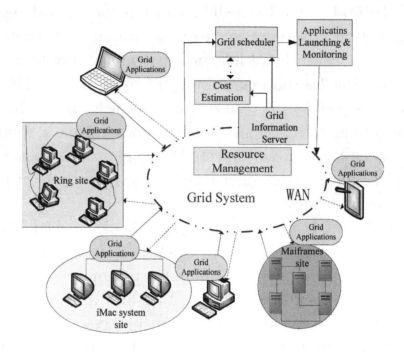

Figure 3.1 Grid scheduling architecture.

Generally, grid systems resources are controlled by the local scheduler that provides access points to resources. Thus, the central grid scheduler must be built on the top of the existing local scheduler [136]. Basically, the central grid scheduler (GS) receives applications from grid users, selects feasible resources for these applications based on the information obtained from the Grid Information Service module, and finally generates application to resource mappings based on a specific objective function and predicted resource performance [43]. The central GS is easier to implement and manage, and is also faster to repair in the event of failure. The role of the grid information service (GIS) is to provide the information about the status of available resources to GSs. Here, GIS can collect and predict the resources status information, such as CPU capacities, communication capacity, memory size, software availabilities, and workload of grid nodes in a particular period [43, 136].

In this grid scheduling architecture, the system is modeled as an undirected graph $GRL=\langle R, L\rangle$, where R is a finite set of p resource vertices, and L is a set of virtual connected undirected arcs or edges [136]. The vertex p_k indicates the grid computing machine k with heterogeneous processing capabilities, such as computation capacity $w(p_k)$ (MIPS), and also includes memory, storages, etc. The edge $l_{n,k}$ denotes a bidirectional communication link between the machines p_n and p_k. The weight $w(l_{n,k})$ of link $l_{n,k}$ stands its data communication capacity [136]. It is also assumed that all inter-machine data communications is performed without link contention, and the communication overhead between two tasks scheduled on the same machine is zero. This assumption holds because the grid computing environment consists of machines connected to the WAN, as pointed out in the literature [70].

This chapter uses the workflow parallel application DAG model as shown in Figure 1.6 (the details can be seen in Section 1.3.2). We assume that both task execution size and edge data communication between tasks in workflow DAG are random, and these are assumed to be independent and exponentially distribution [136]. Table 3.1 lists an example of workflow application DAG task execution size and edge data communication with exponential distribution.

TABLE 3.1 Application task execution size and edge data communication on Figure 1.6

Tasks	t_1	t_2	t_3	t_4	t_5	t_6		
	$f(x, 4.9)$	$f(x, 2)$	$f(x, 3.1)$	$f(x, 3.2)$	$f(x, 1.4)$	$f(x, 2.8)$		
Tasks	t_7	t_8	t_9	t_{10}	t_{11}			
	$f(x, 2.7)$	$f(x, 2.6)$	$f(x, 2.3)$	$f(x, 1.5)$	$f(x, 12.1)$			
Edges	$a_{1,2}$	$a_{1,3}$	$a_{1,4}$	$a_{1,5}$	$a_{2,6}$	$a_{2,9}$	$a_{3,6}$	$a_{3,7}$
	$f(x, 1.7)$	$f(x, 2.2)$	$f(x, 0.3)$	$f(x, 0.5)$	$f(x, 2.1)$	$f(x, 0.1)$	$f(x, 1.1)$	$f(x, 6.9)$
Edges	$a_{4,7}$	$a_{5,6}$	$a_{5,8}$	$a_{6,9}$	$a_{7,10}$	$a_{8,10}$	$a_{9,11}$	$a_{10,11}$
	$f(x, 3.7)$	$f(x, 2.9)$	$f(x, 2.3)$	$f(x, 5)$	$f(x, 2)$	$f(x, 4.1)$	$f(x, 6.2)$	$f(x, 1)$

3.3 STOCHASTIC SCHEDULING PROBLEM

In general, the task execution size and edge communication are deterministic for classical scheduling problem. However, the task execution size and edge communication are random in stochastic scheduling case. Therefore, how to compute the probability of tasks execution time in workflow application graph is the key to stochastic scheduling problem [136]. In this chapter, the execution and communication time are assumed to follow exponentially distributions, which is reasonable to most real-world problems [101, 102, 119, 136]. For ease of understanding, this section first summarizes the symbols used in this chapter in Table 3.2.

TABLE 3.2 Some symbols used in this chapter

Notation	Definition
T	a set of t stochastic tasks in the application
t_i	the ith stochastic task in the application
$w(t_i)$	the probability distribution of task t_i execution size
$a_{i,j}$	the directed edge from ith task to jth task
$w(a_{i,j})$	the probability distribution of edge $a_{i,j}$ data communication
p_k	the kth machine in grid
$w(p_k)$	the kth machine computation capacity
$w(l_{n,k})$	The communication capacity between machines p_n and p_k
$next(t_i)$	a set of immediate successors of task t_i
$prev(t_i)$	a set of immediate predecessors of task t_i
$EES(t_i, p_k)$	the task t_i expected earliest execution start time on machine p_k
$EEF(t_i, p_k)$	the task t_i expected earliest execution finish time on machine p_k
$AWR(Y)$	the approximate weight of random variable Y

3.3.1 The Random Variable Approximate Weight

In classical probability theory, the random variable expected value denotes its average value, which reflects the main attribute of this variable [136]. However, the real-world tasks execution time in computing systems does not equal to their expected value in some case. The other attribute in stochastic mathematical variable is variance, which is the expected value of the square of the deviation. Here, the variance is taking account of all possible probabilities.

The variance attribute in random environment is first considered by work [57], which obtains a good performance approximation. Paper [102] pointed out that the real-world task execution time is mainly affected by the random variable standard deviation. We also prove the computing systems schedule length is bounded by constant Δ, where this phenomenon can be expressed as $Var(Y)/E[Y]^2 \leq \Delta$. For $\Delta \leq 1$, the performances are affected by random variable's expected value and square of variance. Therefore, this chapter takes the random variable variance into account, and the approximate weight of random variable $AWR(Y)$ is defined by [136]

$$AWR(Y) = \begin{cases} E[Y] + \sqrt{Var(Y)} & Var(Y)/E[Y]^2 \leq 1, \\ E[Y]\left(1 + \frac{1}{\sqrt{Var(Y)}}\right) & Otherwise. \end{cases} \tag{3.1}$$

Here, since the tasks execution time and workflow application DAG edge data communication are assumed to follow exponentially distributions, each random variable (such as task execution time, edge communication) are satisfied with $Var(Y)/E[Y]^2 = 1$. Therefore, the approximate weight of random variable is $AWR(Y) = E[Y] + \sqrt{Var(Y)}$. For example, the approximate weight of task t_2 on Table 3.1 is $AWR(t_2) = 0.67$.

3.3.2 Stochastic Scheduling Attributes

This section first defines task t_i approximate earliest execution start time $EES(t_i, p_k)$ and earliest execution finish time $EEF(t_i, p_k)$ with stochastic attribute on machine p_k, which are derived from the deterministic scheduling attributes in Chapter 1, 2. For the entry task t_{entry},

$$EES(t_{entry}, p_k) = 0. \tag{3.2}$$

For the other tasks in workflow stochastic DAG model, the approximate earliest start time and earliest finish time are calculated from the entry task to this task recursively. The equations are shown in Eq. (3.3) and Eq. (3.4), respectively. In this model, all immediate predecessor tasks must have been scheduled so as to obtain the approximate earliest start time of task t_i, which is expressed as

$$EES(t_i, p_k) = \max\{Avail(p_k),$$
$$\max_{t_j \in prev(t_i)} \{EEF(t_j, p_n) + AWR(a_{j,i})\}\}, \qquad (3.3)$$

$$EEF(t_i, p_k) = EES(t_i, p_k) + AWR(t_i)/w(p_k), \qquad (3.4)$$

where $AWR(a_{j,i})$ is the approximate communication stochastic attribute of edge $a_{j,i}$, which is the time of transferring data from immediate predecessor task t_j (scheduled on p_n) to task t_i (scheduled on p_k) and can be computed by Eq. (3.1). Here, it is assumed that $AWR(a_{j,i})$ becomes zero as both t_j and t_i are scheduled on the same machine. The $prev(t_i)$ is the set of immediate predecessor tasks of task t_i, and $Avail(p_k)$ is the expected earliest time at which machine p_k is ready for task execution. After all tasks in workflow application DAG are scheduled, the scheduling objective: schedule length (i.e, overall completion time) is available as the exit task t_{exit} finish time.

Therefore, schedule length (or makespan) is expressed as follows,

$$makespan = EFF(t_{exit}), \qquad (3.5)$$

Here, the $EFF(t_{exit})$ is the task t_{exit} earliest execution finish time, which is different from the approximate earliest finish time $EFF(t_{exit})$. The main objective of workflow tasks scheduling problem is to determine the assignment of tasks to machines such that its schedule length is minimized.

3.4 THE STOCHASTIC SCHEDULING STRATEGY

This section presents a Stochastic Heterogeneous Earliest Finish Time scheduling algorithm (SHEFT) on grid computing systems, which aims to obtain high scheduling performance and low complexity [136]. The stochastic scheduling algorithm mainly consists of two schemes: the listing scheme that is a modified version of the classical heuristic HEFT scheduling algorithm [146], and machine assignment scheme that each task in workflow DAG is assigned to the corresponding machine to minimize the schedule length [136]. The pseudo-code of SHEFT algorithm is shown in Algorithm 1. This section first outlines the basic concept of workflow DAG stochastic task priority. Then, the stochastic list scheduling algorithm is presented and analyzed.

Input: The grid application tasks and grid systems GRL
Output: The scheduling of task-machine pairs
1 Compute $SRank$ value for all tasks using Eq. (3.6) by traversing application graph;
2 Sort the scheduling list tasks by non-increasing order of $SRank$ value;
3 **while** *there are unscheduled tasks in the list* **do**
4   Remove the first task t_i from the scheduling list;
5   **for** *each machine $p_k \in P$* **do**
6   Compute the approximate earliest start time using Eq. (3.3);
7   Compute the approximate earliest finish time using Eq. (3.4);
8   **end**
9   Find the minimum approximate earliest finish time machine p_n;
10   Assign task t_i to machine p_n with minimize $EEF(t_i, p_n)$;
11 **end**

Algorithm 1: The SHEFT algorithm

3.4.1 Stochastic Task Priorities Phase

This section first defines the stochastic upward rank ($SRank$), which is used to compute tasks priorities in the stochastic heterogeneous earliest finish time scheduling algorithm [136]. The $SRank$ is defined in Definition 3.1.

Definition 3.1 *Given a workflow parallel application DAG with stochastic execution and communication time, and grid computing systems with m machines. The SRank during a particular scheduling step is a rank of task, from an exit task to itself, which* *has the sum of approximate task execution time and approximate data communication between tasks over all grid machines.*

The *SRank* is recursively expressed as follows,

$$
SRank(t_i) = \frac{1}{m} \sum_{k=1}^{m} \frac{AWR(t_i)}{w(p_k)} + \\
\max_{t_j \in next(t_i)} \{ AWR(e_{i,j})/\overline{w(l)} + SRank(t_j) \}. \tag{3.6}
$$

Here, $next(t_i)$ denotes the set of immediate successor tasks of t_i, the approach $(\sum_{k=1}^{m} AWR(t_i)/w(p_k))/m$ is an approximate mean stochastic rank of task t_i that take task t_i expected value and variance into account, $\overline{w(l)}$ is the mean communication capacity on grid computing systems [136]. The *SRank* is computed recursively by traversing the workflow DAG task upward, starting from the exit task. In this section, the *SRank* of exit task t_{exit} is equal to

$$
SRank(t_{exit}) = \frac{1}{m} \sum_{k=1}^{m} \frac{AWR(t_{exit})}{w(p_k)}. \tag{3.7}
$$

Basically, $SRank(t_i)$ is the length of the workflow DAG stochastic critical path from the current task t_i to exit task, including the approximate mean stochastic rank of itself [136].

3.4.2 Machine Selection Phase

In the phase of machine selection and task assignment, the unscheduled task is selected and scheduled on a grid machine that can obtain the minimum execution time. To achieve this goal, the SHEFT algorithm arranges task scheduling sequence according to the task priority *SRank*, which takes the attributes of workflow task random into account. For each DAG model stochastic task, the SHEFT use Eq. (3.3) and Eq. (3.4) to compute $EEF(t_i, p_n)$, which can be implemented from step 5 to 8 [136]. In step 9, SHEFT algorithm try to find a machine with minimize $EEF(t_i, p_n)$ and assigns task t_i to the corresponding machine p_n in step 10. It is to be noted that this machine may or may not be its best-suited machine due to the workflow DAG task stochastic [136].

3.4.3 SHEFT Scheduling Algorithm Complexity Analysis

The time-complexity of list scheduling algorithms for high-performance worklfow application is usually expressed as the number of task T, number of edges A, and the number of target grid computing systems machines m. Here, the time-complexity of SHEFT algorithm is analyzed as follows: Computing the *SRank* of workflow DAG task can be done in time $max\{O(|T||A|), O(|T|m)\}$, Sorting the tasks can be done in time $O(|T|\log|T|)$ [136]. In the machine selection phase, finding the optimal machine can be done in time $O(\Omega|T|m)$, and Ω is the max degree of t_i in DAG graph [136]. In grid computing systems, the number of machines may larger than the number of parallel application DAG edges in most case. Therefore, the time-complexity of the stochastic heterogeneous earliest finish time scheduling algorithm is $O(\Omega|T|m)$ [136].

3.5 ALGORITHM PERFORMANCE EVALUATION

This section compares the performance of stochastic heterogeneous earliest finish time scheduling algorithm (SHEFT) with two well-known existing classical algorithms on grid computing systems: the HEFT (the details can be seen in Section 2.4.3) and DCP algorithms [146]. To make the comparison efficient, we transform the stochastic scheduling problem into deterministic scheduling, and use the expected value as the task weight in HEFT and DCP algorithms. The purpose of performance comparison is not only to give quantitative results, but also to qualitatively analyze and explain the experimental results, so as to better understand the whole scheduling problem [136].

3.5.1 Experiments Setting and Evaluation Metrics

This chapter conducts a simulation platform of grid computing systems with 16 machines, where their computation capacity, memory, and storage vary from Pentium II to Pentium IV. The wide area grid communication network are randomly generated and communication bandwidth are assumed to be uniformly distributed between 10 and 100 Mbits/sec [136].

The algorithm performance evaluation metrics chosen in this chapter are speedup, schedule length (or makespan Eq. (3.5)), and makespan standard deviation [136]. The speedup is a classical parallel algorithm evaluation metric that the sequential execution time divides the schedule length of parallel application, which is defined by

$$speedup = \frac{\sum_{t_i \in T} AET(t_i)}{makespan}, \quad (3.8)$$

where $AET(t_i)$ is the actual execution time of workflow application task. The sequential execution time is computed by assigning all tasks to a single machine [136].

This chapter introduces a performance evaluation metric: makespan standard deviation [136]. Intuitively, the standard deviation of the scheduling performance makespan distribution tells how about the makespan distribution. The centralized distribution means the small value of the makespan standard deviation. The makespan standard deviation is examined in this experiments because when you are given two scheduling solutions, the one with smaller standard deviation is the one that is more likely to have stable performance on grid computing systems [136].

3.5.2 Randomly Generated Workflow DAG Graphs

This chapter considers the randomly generated stochastic workflow parallel application DAG graphs. Here, we implement a randomly workflow application DAG graph generator to generate stochastic DAGs with various characteristics that depend on serval input parameters, such as *DAG size, Height of the DAG* [136].

The grid scheduling framework allows assigning values to the parameters by the generator. The framework first uses the random workflow DAG graph generator to construct the stochastic parallel application DAG tasks [136]. Then, the scheduling algorithms are provide to produce the grid systems scheduling solutions. Finally, we compute the performance metrics according to the tasks schedules. For the generation of stochastic parallel application DAG graphs, three fundamental characteristics of workflow DAG are considered:

- *DAG size*: The number of tasks in the workflow parallel application DAG.

- *Height of the DAG(h)*: The DAG tasks are randomly partitioned into h levels.

- The maximum and minimum λ value $(\lambda_{max}, \lambda_{min})$ of task execution size and data communication between tasks with exponentially distribution: The λ value is a uniform random variable on the interval $(\lambda_{max}, \lambda_{min})$.

In these experiments, the simulated application DAGs are generated by the combinations of the above parameters with the number of DAG tasks between 50 and 300 [136]. Each edge is generated with the same probability, which is computed by the average number of DAG edges per DAG task. Every set of the above parameters are used to generate several workflow DAG graphs in order to obtain good application DAG. The experimental results presented below are the average of these workflow application DAG graphs [136].

3.5.3 The Sensitivity of Machine Number

In the first simulation experiments, we compare three algorithms with the sensitivity of machine number, which varies from 4 to 16 in steps of 2. The experimental results are shown in Figures 3.2, 3.3, and 3.4 for 100, 200, 300 tasks, respectively. For the performance evaluation metrics: schedule length and speedup, each data is the average obtained in 100 experiments, and the makespan standard deviation metric, data are the distribution of these 100 experiments [136].

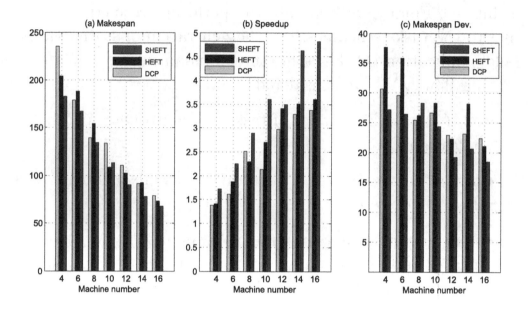

Figure 3.2 Experimental results of 100 tasks. (a) makespan; (b) speedup; (c) makespan standard deviation.

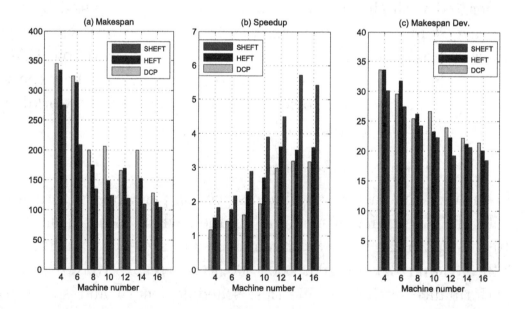

Figure 3.3 Experimental results of 200 tasks. (a) makespan; (b) speedup; (c) makespan standard deviation.

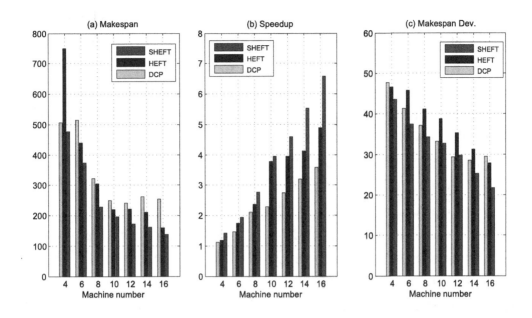

Figure 3.4 Experimental results of 300 tasks. (a) makespan; (b) speedup; (c) makespan standard deviation.

Figure 3.2 shows the results of SHEFT, DCP, and HEFT algorithms on grid computing systems for application DAG with 100 tasks. From Figure 3.2, we can conclude that the SHEFT algorithm outperforms DCP and HEFT by 16.1% and 10.7%, respectively, in term of the average schedule length or makespan. The SHEFT algorithm is also outperforms DCP, HEFT by 26.1% and 19.7% in term of the average speedup, and 9.97%, 21.3% in term of makespan standard deviation, respectively [136].

This is mainly due to the fact that the stochastic heterogeneous earliest finish time scheduling algorithm considers the parallel application task execution stochastic attribute, such as expected value and variance, which produces a good schedule for stochastic task scheduling problem $Q|t_i \sim stoch, prec|E[C_{max}]$ [136]. However, the deterministic DCP and HEFT algorithms ignore the available stochastic information about task execution and data communication between tasks, which assign machine for tasks with the expected value of random variables. For instance, the task execution time follows exponentially distribution as $f(x) = 0.1e^{-0.1x}$, the expected value is 10 and the variance is 100 [136]. In this case, the scheduling decisions of DCP and HEFT are according to the expected value 10, which is far from the task actual execution. However, for the SHEFT algorithm, the scheduling decision is made according to the task value of 20, which is close to the actual execution time [136].

By comparing simulation results of HEFT and DCP algorithms, we can find that HEFT is better than DCP in terms of speedup and schedule length. This is mainly due to the fact that classical HEFT algorithm can more effectively take advantage of workflow application DAG critical tasks than DCP [136]. However, for the evaluation metric makespan standard deviation, DCP is better than HEFT. This phenomena can be attributed to the fact that the variance of DCP over HEFT [136]. As the grid computing systems machine number increases, the makespan and makespan standard deviation of three algorithms decreases, and the speedup of them increases.

Figure 3.3 shows the results of application with 200 tasks. From Figure 3.3, we can conclude that the SHEFT outperforms DCP by 46.2%, 49.96%, and 12.5%, HEFT by 30.7% and 39.05% and 9.8%, in terms of the average makespan, average speedup, and makespan standard deviation, respectively [136]. However, the improvements of the SHEFT over DCP, HEFT are almost at the same level except that the machine number is 14, which reflect the intricacy of stochastic scheduling problem. The improvements of SHEFT over DCP, HEFT could be also observed from Figure 3.4, which shows the results of workflow DAG with 300 tasks. The SHEFT algorithm is better than DCP by 34.3%, 29.1%, and 9.7%, HEFT by 32%, 21.6%, and 18.6%, in terms of average makespan, speedup, and makespan standard deviation, respectively [136].

3.5.4 The Sensitivity of DAG Size

The second set of experiments try to examine the performance evaluation metric sensitivity of the algorithms SHEFT, HEFT

TABLE 3.3 Performance impact of 10 machines for makespan

Tasks	50	100	150	200	250	300
SHEFT	34.33	66.46	114.49	123.66	168.02	196.65
HEFT	47.83	87.39	140.17	148.78	208.09	219.41
DCP	51.65	88.26	118.44	206.12	194.1	248.3

TABLE 3.4 Performance impact of 10 machines for speedup

Tasks	50	100	150	200	250	300
SHEFT	1.68	2.09	3.57	3.91	4.64	4.95
HEFT	1.54	1.9	2.35	3.7	3.56	3.78
DCP	1.53	1.86	2.92	2.93	2.95	2.27

and DCP to the workflow DAG size. Here, we vary grid parallel application DAG size from 50 to 300 in steps of 50. The experimental results reported in Tables 3.3, 3.4, and 3.5 are the makespan, speedup, makespan standard deviation of 10 machines, respectively [136]. From these tables, we can conclude that the SHEFT outperforms HEFT by 21%, DCP by 28.8 in terms of average makespan. The SHEFT is also better than HEFT by 19.2%, DCP by 30.6% in terms of average speedup, and the SHEFT is over HEFT by 12.2%, DCP by 13.6% in terms of average makespan standard deviation.

TABLE 3.5 Performance impact of 10 machines for makespan standard deviation

Tasks	50	100	150	200	250	300
SHEFT	20.9	24.34	23.87	22.34	28.74	32.6
HEFT	23.97	28.23	26.16	23.23	31.17	38.78
DCP	24.19	26.6	27.7	26.6	35.4	33.2

Tables 3.6, 3.7, and 3.8 are the experimental results on 16 machines [136]. From these Tables, we can draw a similar conclusion that SHEFT outperforms HEFT, DCP in terms of makespan, speedup, and makespan standard deviation. Moreover, as the application DAG size increases, the improvement

becomes more significant. The above simulation results also show a fact that the deterministic scheduling algorithm (such as HEFT, DCP) is not suitable for stochastic scheduling problem $Q|t_i \sim stoch, prec|E[C_{max}]$.

TABLE 3.6 Performance impact of 16 machines for makespan

Tasks	50	100	150	200	250	300
SHEFT	27.35	63.92	101.61	104.53	146.86	137.74
HEFT	27.86	63.91	119.29	113.2	165.43	159.64
DCP	47.32	85.33	110.18	128.6	229.83	252.9

TABLE 3.7 Performance impact of 16 machines for speedup

Tasks	50	100	150	200	250	300
SHEFT	2	2.28	3.41	5.42	5.9	6.57
HEFT	2.18	2.43	2.82	3.6	4.26	4.88
DCP	1.6	2.5	2.56	3.17	3.77	3.58

TABLE 3.8 Performance impact of 16 machines for makespan standard deviation

Tasks	50	100	150	200	250	300
SHEFT	18.2	18.47	19.2	18.47	19.4	23.7
HEFT	19.3	21.1	23.6	20.1	24.7	27.8
DCP	21.3	22.43	23.7	21.43	25.4	29.43

3.6 SUMMARY

This chapter first describes the grid scheduling architecture and formulate the stochastic workflow parallel application DAG model. Then, we presents a stochastic heterogeneous earliest finish time scheduling algorithm to solve the grid task scheduling problem. The algorithm is derived from the deterministic scheduling algorithm and combines the stochastic attribute of task execution time and data communication into grid task scheduling.

CHAPTER 4

Scheduling Stochastic Tasks on Heterogeneous Cluster Systems

I N general, workflow parallel application consists of precedence constrained stochastic tasks, where the task execution time and inter-task data communication are random variables that follow some probability distribution. In the parallel and distributed computing systems, scheduling precedence constrained stochastic tasks and minimizing the expected completion time of parallel applications on heterogeneous cluster system with processors with varying computing power is a very difficult but important problem. This chapter presents stochastic scheduling attributes, methods dealing with various random variables, and a method of scheduling stochastic parallel applications on cluster systems. At the same time, in order to understand how to solve the scheduling problem of precedence constrained stochastic tasks effectively and efficiently, we also present a stochastic dynamic level scheduling (SDLS) algorithm based on stochastic bottom levels and stochastic dynamic levels.

DOI: 10.1201/b23006-4

4.1 INTRODUCTION

In the recent years, cluster systems have become primary and cost-effective high-performance computing infrastructures for large-scale parallel processing. Parallel processing is a promising technology to meet the computational requirements of a great deal of current and emerging parallel applications, including neural network training, weather modeling, information processing, image processing, and fluid flow [81]. The computations and data of these applications can be distributed on the cluster systems processors, and the benefits can be obtained from such systems by adopting efficient job partitioning and scheduling approaches [81]. More and more evidence show that scheduling parallel workflow applications is very important to the performance of the cluster systems. The common goal of resource management and task scheduling is to map workflow application tasks onto cluster systems processors, and their executions order is arranged to meet the task precedence constraints and achieve the minimum schedule length or makespan [21, 71, 120, 168].

The traditional scheduling research work and algorithms mainly focus on the deterministic task scheduling problem with certain computation and communication time (see Chapters 2 and 3) [21, 71, 75, 90, 120, 137, 168]. However, in real-world execution environments, tasks may not have deterministic execution time. In general, parallel application tasks usually contain conditional instructions, which may produce different execution time with different inputs [95, 97, 105, 145]. For example, the smoothed particle hydrodynamics computation in interfacial flows numerical simulation [30]. Furthermore, data communication time between tasks can also fluctuate according to network traffic [81].

Although the deterministic scheduling methods can thoroughly find the best assignment solution of tasks to processors, existing scheduling algorithms are not effectively dealt with the uncertainty of parallel applications [81]. A natural way to overcome this difficulty is to consider task scheduling with stochastic. That is, to interpret task execution time and data communication between tasks as random variables, and to measure the performance of scheduling strategy by its expected objective value.

Here, we let $T = \{t_1, t_2, \ldots, t_n\}$ denote a set of n tasks in workflow application DAG model, which need to be scheduled on a cluster computing systems with m heterogeneous processors, so as to minimize the schedule length [81]. In contrast to deterministic task scheduling, the important assumption of stochastic tasks is that the task execution size $w(t_i)$ of t_i is not fixed and known in advance. Instead, this chapter assumes that the task execution size $w(t_i)$ is a random variable, where we just give their probability distribution function [81]. Similarly, the inter-task data communication time is also assumed like this.

There are some study works on stochastic task scheduling problem in recent years [23, 42, 114, 128, 136]. In paper [42], Dong et al. proposed a scheme for estimating the probability distribution of task execution time based on computing resource workload and developed a grid systems stochastic task scheduling algorithm. The weakness of this approach is that the task probability distribution is far from the real-world and inefficient. There are remain some challenging problems, such as, how to incorporate the expected values and variances of random time into scheduling to improve the cluster systems performance, and how to deal with processor heterogeneity in cluster computing systems [81]. These issues are worth of further study and there is still much room for improvement [97, 128].

4.2 A STOCHASTIC SCHEDULING MODEL

4.2.1 Stochastic Workflow Applications

In general, a stochastic workflow parallel application with precedence constrained tasks is represented by a directed acyclic graph (DAG) $G = \langle T, A \rangle$ [136, 146](the details can be seen in

Section 1.3.2). In this stochastic scheduling architecture, the workflow DAG task execution and edge communication time are random and known only in advance as certain probability distributions [81].

This chapter presents the stochastic workflow application task scheduling problem that task execution and edge communication time with normal distributions, and it is also assumed that the expected values and variances are known in advance [81]. The assumption of normal distribution has been proved to be correct by many practical applications. That approach has been adopted by many researchers [7, 97, 118]. Here, we let $N(\mu, \sigma^2)$ is the normal probability distribution with mean μ and variance σ^2. Therefore, the task t_i execution time follows $w(t_i) \sim N(\mu_{t_i}, \sigma^2_{t_i})$, and the edge $a_{i,j}$ data communication time follows $w(a_{i,j}) \sim N(\mu_{a_{i,j}}, \sigma^2_{a_{i,j}})$, respectively [81].

Figure 4.1 shows an example of a workflow parallel application, where its task processing and edge edge communication time follow a normal distribution.

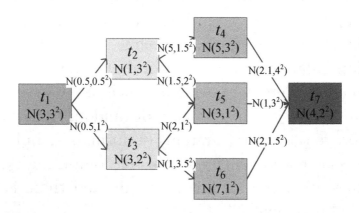

Figure 4.1 A workflow application with normal distribution

4.2.2 Heterogeneous Cluster Systems

This chapter considers a cluster computing systems that is modelled as a finite set $P = \langle p_1, p_2, ..., p_m \rangle$ with m heterogeneous processors [81]. A weight $w(p_k)$ assigned to a processor p_k denotes the processor's computation capacity. Therefore, the task t_i on processor p_k processing time can be defined as $w(t_i)/w(p_k)$. In this systems, all processors are connected by a special interconnection network, such as InfiniBand. We further assume that the data communication time between tasks t_i and t_j as $w(a_{i,j})$. If the two tasks are executed on different processors, the time is following normal distribution [81]. However, as the two tasks scheduled on the same processor, the time is taken as zero [75, 146].

4.2.3 The Motivational Example

In order to demonstrate the complexity of scheduling random tasks (only its probability distribution is known in advance) and minimize the schedule length or makespan, we give an illustration of the workflow DAG task execution and edge communication time in the Figure 4.1. Table 4.1 shows the average-case execution time and the worst-case execution time of the tasks in Figure 4.1, which are used by traditional deterministic scheduling algorithms [81]. Table 4.2 gives the processing time of the tasks in Figure 4.1, which are obtained from a cluster computing systems based on Intel Xeon X5670 processors and Linux system. We observe from Table 4.2 that the task execution time or an edge data communication time can be far from its average-case or worst-case [81]. Therefore, the deterministic scheduling scheme is ineffective and we should use the stochastic scheduling strategy to improve the task scheduling performance. The example will be continued and completed in later sections.

4.3 THE PRELIMINARY CONCEPTS

For ease of reference, this section summarizes the notations and definitions used in this chapter in Table 4.3.

TABLE 4.1 The deterministic scheduling of Figure 4.1

Node	t_1	t_2	t_3	t_4	t_5	t_6	t_7		
Average-case	3.0	1.0	3.0	5.0	3.0	7.0	4.0		
Worst-case	7.3	2.5	5.6	12.1	4.8	10.1	6.9		
Edge	$a_{1,2}$	$a_{1,3}$	$a_{2,4}$	$a_{2,5}$	$a_{3,5}$	$a_{3,6}$	$a_{4,7}$	$a_{5,7}$	$a_{6,7}$
Average-case	0.50	0.5	5.0	1.5	2.0	1.0	2.1	1.0	2.0
Worst-case	0.83	1.2	8.9	3.2	2.8	2.1	5.3	2.2	3.6

TABLE 4.2 The processing time of Figure 4.1 on cluster systems

Number	t_1	t_2	t_3	t_4	t_5	t_6	t_7		
1	4.5	0.70	2.3	12.0	3.9	5.6	2.90		
2	3.3	2.10	2.9	8.1	3.4	9.1	6.40		
3	3.2	0.89	3.6	3.5	3.3	9.3	3.12		
Number	$a_{1,2}$	$a_{1,3}$	$a_{2,4}$	$a_{2,5}$	$a_{3,5}$	$a_{3,6}$	$a_{4,7}$	$a_{5,7}$	$a_{6,7}$
1	1.50	0.20	7.3	0.90	2.60	0.71	3.22	1.43	1.21
2	0.81	0.30	3.2	2.82	1.80	1.32	4.13	1.92	2.16
3	0.56	0.80	6.7	1.89	2.73	1.90	2.30	0.90	2.80

TABLE 4.3 Notations and definitions

Notation	Definition
T	a set of n tasks in stochastic workflow application
t_i	the ith stochastic task
$w(t_i)$	a random variable denoting the execution size of task t_i
A	a set of edges denoting data communication among tasks in T
$a_{i,j}$	the edge denoting data communication from task t_i to task t_j
$w(a_{i,j})$	random variable denoting data communication of edge $a_{i,j}$
P	a set of m heterogeneous processors
p_k	the kth processor in P
$w(p_k)$	the computation capacity of processor p_k
$next(t_i)$	the set of immediate successors of task t_i
$prev(t_i)$	the set of immediate predecessors of task t_i
$prc(t_i)$	the processor on which task t_i is executed
$EES(t_i, p_k)$	the earliest execution start time of task t_i on processor p_k
$AET(t_i, p_k)$	the execution time of task t_i on processor p_k
$EEC(t_i, p_k)$	the earliest execution finish time of task t_i on processor p_k
$FTP(p_k)$	the finish time of processor p_k
$EEC(a_{j,i})$	the communication finish time of edge $a_{j,i}$
$RT(t_i, p_k)$	the data ready time of task t_i on processor p_k

4.3.1   Scheduling Attributes

To describe the scheduling problem of workflow parallel application DAG $G = \langle T, A \rangle$ on heterogeneous cluster computing systems, the following notations are defined and used throughout this chapter [81].

We let $EES(t_i, p_k)$ denote the earliest execution start time of task $t_i \in T$ on processor $p_k \in P$, and $AET(t_i, p_k){=}w(t_i)/w(p_k)$ is the actual execution time of task t_i on processor p_k [81]. As the computing capacities of processors in a heterogeneous systems are different from each other, the execution times, i.e., the $AET(t_i, p_k)$'s, are different too [81]. The earliest execution completion time $EEC(t_i, p_k)$ of task t_i on processor p_k is computed by,

$$
\begin{aligned}
EEC(t_i, p_k) &= EES(t_i, p_k) + AET(t_i, p_k) \\
&= EES(t_i, p_k) + w(t_i)/w(p_k).
\end{aligned}
\tag{4.1}
$$

Notice that in this chapter, we assume that the task execution size $w(t_i)$ is a random variable with normal distribution. Therefore, the actual execution time $AET(t_i, p_k)$ and the earliest completion time $EEC(t_i, p_k)$ are random too [81].

The processor on which task t_i assigned is expressed as $prc(t_i)$. Thus, a schedule solution $s = (prc(t_1), prc(t_2), ..., prc(t_n))$ is a task assignment to the heterogeneous cluster systems processors (equivalently, a processor allocation to the tasks) [81]. Moreover, we let $FTP(p_k){=}MAX_{prc(t_i)=p_k}\{EEC(t_i, p_k)\}$ be the finish time of processor p_k. For a schedule solution s to be feasible, the following condition must be satisfied by all tasks in the workflow application DAG [125], i.e., for any two tasks $t_i, t_j \in T$,

$$
prc(t_i) = prc(t_j) = p_k \Rightarrow
\begin{cases}
EEC(t_i, p_k) \leq EES(t_j, p_k), \\
or \\
EEC(t_j, p_k) \leq EES(t_i, p_k).
\end{cases}
$$

The above condition essentially means that two tasks executed on the same processor cannot overlap [81]. The data ready time (RT) of task t_i on processor p_k is denoted as

$$RT(t_i, p_k) = MAX_{t_j \in prev(t_i), prc(t_j) \neq p_k} \{EEC(a_{j,i})\}, \qquad (4.2)$$

where $EEC(a_{j,i})$ is the communication finish time of edge $a_{j,i}$ and is defined by,

$$EEC(a_{j,i}) = EEC(t_j, p_{prc(t_j)}) + w(a_{j,i}). \qquad (4.3)$$

If $prev(t_i) = \emptyset$, i.e., t_i is an entry task, then we let $RT(t_i, p_k)$, for all $p_k \in P$. Here, the task t_i on processor p_k start time $EES(t_i, p_k)$ is constrained by processor p_k execution finish time $FTP(p_k)$ and edges communication as the following [81],

$$EES(t_i, p_k) = MAX\{RT(t_i, p_k), FTP(p_k)\}, \qquad (4.4)$$

for all $t_i \in T$ and $p_k \in P$.

The main scheduling goal of makespan is $EEC_{max} = EEC(t_{exit}, p_{prc(t_{exit})})$. In this solutions, it is assumed that $EES(t_{entry}, p_{prc(t_{entry})}) = 0$. In solving the $Q|t_i \sim stoch, prec|E[EEC_{max}]$ problem, one of the objectives is to minimize the average makespan $E[EEC(t_{exit}, p_{prc(t_{exit})})]$ [81].

4.3.2 Manipulation of Normal Random Variables

Since the task execution and data communication time in a stochastic workflow application are given as random variables, many quantities in stochastic scheduling problem, such as the length of critical path from a task to the *exit* task are also random variables, which differ far from the deterministic problems [81]. Before presenting the workflow DAG task stochastic dynamic level scheduling (SDLS) algorithm for the $Q|t_i \sim stoch, prec|E[EEC_{max}]$ problem, we would like to explain how to compute the normal probability distributions of mathematic random variables involved in stochastic scheduling problem.

There are several basic operations on normal distributions which are adopted throughout this chapter. First, a normal distribution random variable $X \sim N(\mu, \sigma^2)$ can be scaled by a constant weight b, e.g., the task execution random variable $w(t_i)$ is divided by a processor computation capacity $w(p_k)$ in Eq. (4.1) [81]. It is clear that $bX \sim N(b\mu, (|b|\sigma)^2)$.

Secondly, we encounter summations of normal distribution variables. The important fact about normal probability distribution is that if X_i has a normal distribution with expected value μ_i and variance σ_i^2, where $i = 1, 2, \ldots, n$, then $X = \Sigma_i^n X_i$ also has a normal distribution with parameters $\Sigma_i^n \mu_i$ and $\Sigma_i^n \sigma_i^2$ [81]. In word, X is defined as follows:

$$X \sim N(\Sigma_i^n \mu_i, \Sigma_i^n \sigma_i^2). \tag{4.5}$$

This attribute can be adopted to compute the distribution of the tasks path length in a workflow application DAG, such as Figure 4.2(a). Here, the path length normal distribution $(t_x \rightarrow a_{x,y} \rightarrow t_y)$ in Figure 4.2(a) is $N(\mu_x + \mu_{x,y} + \mu_y, \sigma_x^2 + \sigma_{x,y}^2 + \sigma_y^2)$. We also notice that $X - Y = X + (-Y)$, i.e., the difference between two normal random variables is also following normal distribution [81]. This property will be used in later Eq. (4.13).

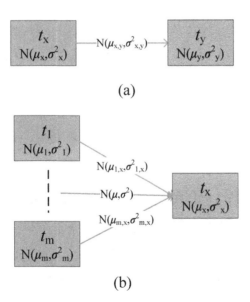

Figure 4.2 Stochastic workflow DAG series and parallel model.

Thirdly, we need to calculate the maximum of several normal distribution variables, such as $RT(t_i, p_k)$ in Eq. (4.2) and the MAX operator in Eq. (4.4). Unfortunately, the maximum probability of following normal distribution is no longer a normal distribution random variable [81]. However, in the pioneering work [118], [33], [28], Clark proposed an approach to recursively approximate estimate the expected value and variance of the maximum value of a finite group of random variables with normal distribution. This distribution is close to the actual distribution of the MAX operator. Here, we consider the tasks in Figure 4.2(b), where task t_x has m predecessors. The task t_x data ready time can be computed by

$$
\begin{aligned}
RT(t_x) &= MAX\{EEC(a_{1,x}), \cdots, EEC(a_{m,x})\} \\
&= MAX\{MAX\{EEC(a_{1,x}), \cdots, EEC(a_{m,x})\}.
\end{aligned}
\tag{4.6}
$$

The above equation means that the probability distribution of $RT(t_x)$ can be computed recursively by using Clark's equations [81]. As the data communication is independent and follow normal distribution, the correlation coefficient between any pair of them is zero [81]. That is,

$$
\rho_{i,j} = \rho(EEC(a_{i,n}), EEC(a_{j,n})) = 0, \quad \forall i, j = 1, 2, ..., n-1,
$$
$$
and \ i \neq j.
$$

Here, we use Clark's equations to compute the expected value and variance of $MAX\{EEC(a_{1,n}), EEC(a_{2,n})\}$ with $\rho_{1,2} = 0$, which are as follows [81]. The expectation of $MAX\{EEC(a_{1,n}), EEC(a_{2,n})\}$ is computed by Clark's first equation. That is,

$$
\begin{aligned}
E[MAX\{EEC(_{1,n}), EEC(a_{2,n})\}] &= E[EEC(a_{1,n})]\Phi(\xi_{1,2}) \\
&+ E[EEC(a_{2,n})]\Phi(-\xi_{1,2}) + \varepsilon_{1,2}\psi(\xi_{1,2}) \\
&= E[EEC(a_{2,n})] + (E[EEC(a_{1,n})] - \\
&E[EEC(a_{2,n})])\Phi(\xi_{1,2}) + \varepsilon_{1,2}\psi(\xi_{1,2}),
\end{aligned}
\tag{4.7}
$$

where,

$$
\begin{aligned}
\varepsilon_{1,2} &= \sqrt{Var[EEC(a_{1,n})] + Var[EEC(a_{2,n})] - 2\rho_{1,2}\chi_{1,2}} \\
&= \sqrt{Var[EEC(a_{1,n})] + Var[EEC(a_{2,n})]},
\end{aligned}
$$

(since $\rho_{1,2} = 0$), and

$$\xi_{1,2} = \frac{E[EEC(a_{1,n})] - E[EEC(a_{2,n})]}{\varepsilon_{1,2}},$$

and $\chi_{1,2} = \sqrt{Var[EEC(a_{1,n})]Var[EEC(a_{2,n})]}$, and

$$\psi(x) = \frac{1}{\sqrt{2\pi}}e^{-\frac{1}{2}x^2},$$

and

$$\Phi(y) = \int_{-\infty}^{y} \psi(x)dx = \frac{1}{\sqrt{2\pi}} \int_{-\infty}^{y} e^{-\frac{1}{2}x^2} dx.$$

The variance of $MAX\{EEC(a_{1,n}), EEC(a_{2,n})\}$ is calculated by Clark's second equation. That is,

$$Var[MAX\{EEC(a_{1,n}), EEC(a_{2,n})\}] = (E^2[EEC(a_{1,n})] +$$
$$Var[EEC(a_{1,n})])\Phi(\xi_{1,2}) + (E^2[EEC(a_{2,n})] + Var[EEC(a_{2,n})])$$
$$\Phi(-\xi_{1,2}) + (E[EEC(a_{1,n})] + E[EEC(a_{2,n})])\varepsilon_{1,2}\psi(\xi_{1,2})$$
$$- E^2[MAX\{EEC(a_{1,n}), EEC(a_{2,n})\}].$$

$$(4.8)$$

Now, let us consider,

$$MAX\{EEC(a_{1,n}), EEC(a_{2,n}), EEC(a_{3,n})\}$$
$$= MAX\{MAX\{EEC(a_{1,n}), EEC(a_{2,n})\}, EEC(a_{3,n})\}.$$

The correlation coefficient $\rho_{1,2,3}$ becomes the correlation coefficient between $MAX\{EEC(a_{1,n}), EEC(a_{2,n})\}$ and $EEC(a_{3,n})$. It is clear that $\rho_{1,2,3} = 0$. The expected value of $MAX\{EEC(a_{1,n}), EEC(a_{2,n}), EEC(a_{3,n})\}$ is defined as

$$E[MAX\{EEC(a_{1,n}), EEC(a_{2,n}), EEC(a_{3,n})\}] = E[MAX\{$$
$$EEC(a_{1,n}), EEC(a_{2,n})\}]\Phi(\xi_{1,2,3}) + E[EEC(a_{3,n})]\Phi(-\xi_{1,2,3})$$
$$+ \varepsilon_{1,2,3}\psi(\xi_{1,2,3}),$$

where,

$$\varepsilon_{1,2,3} = \sqrt{Var[MAX\{EEC(a_{1,n}), EEC(a_{2,n})\}] + Var[EEC(a_{3,n})]},$$

and

$$\xi_{1,2,3} = \frac{E[MAX\{EEC(a_{1,n}), EEC(a_{2,n})\}] - E[EEC(a_{3,n})]}{\varepsilon_{1,2,3}}.$$

The variance of $MAX\{EEC(a_{1,n}), EEC(a_{2,n}), EEC(a_{3,n})\}$ can be calculated by,

$$Var[MAX\{EEC(a_{1,n}), EEC(a_{2,n}), EEC(a_{3,n})\}]$$
$$= (E^2[MAX\{EEC(a_{1,n}), EEC(a_{2,n})\}] + Var[MAX\{EEC(a_{1,n}),$$
$$EEC(a_{2,n})\}])\Phi(\xi_{1,2,3}) + (E^2[EEC(a_{3,n})] + Var[EEC(a_{3,n})])$$
$$\Phi(-\xi_{1,2,3}) + (E[MAX\{EEC(a_{1,n}), EEC(a_{2,n})\}] +$$
$$E[EEC(a_{3,n})])\varepsilon_{1,2,3}\psi(\xi_{1,2,3}) - E^2[MAX\{EEC(a_{1,n}),$$
$$EEC(a_{2,n}), EEC(a_{3,n})\}].$$

Using a similar approach, the Clark equations can be recursively used for $(n-3)$ more steps to calculate the expected value and variance of task t_n data ready time $RT(t_n)$ [81].

4.4   A STOCHASTIC SCHEDULING ALGORITHM

4.4.1   Stochastic Bottom Level

In the heterogeneous cluster computing systems, parallel application tasks executed on different processors with different execution time give us a problem. That is, the priorities calculated using the task execution time on one specific processor may be different from that of other processor [81]. To overcome this difficulty, some scheduling algorithms let the task execution time as their median value, such as DLS algorithm in the paper [120]. And, some algorithms use their average value, such as HEFT algorithm in the [146]. For heterogeneous computing systems, the impacts of various methods are investigated in [174], where they compared different schemes such as median value, best value, average value, and worst value. The experimental results show that the average value is the most suitable approach for heterogeneous systems. Therefore, this chapter uses the systems average computation capacity [81], which is expressed as

$$\overline{w(p)} = \frac{1}{m} \sum_{k=1}^{m} w(p_k). \tag{4.9}$$

In deterministic task scheduling techniques, the important attributes such as bottom level *blevel* and top level *tlevel* can be easily calculated by approaches proposed in the papers [75]. However, for stochastic task scheduling strategies, it is difficulty to calculate such attributes, because the tasks execution and data communication are random [81]. This section presents a method to compute the probability distribution of *blevel*, where we call it as *stochastic bottom level* and expressed as *sblevel* [81]. Generally, the *sblevel* of task t_x is the random distribution of the critical or longest path from task t_x to the *exit* task [81], and can be recursively computed as follows,

$$\begin{aligned} sblevel(t_x) = &MAX_{t_i \in next(t_x)}\{w(a_{x,i}) \\ &+ sblevel(t_i)\} + w(t_x)/\overline{w(p)}. \end{aligned} \tag{4.10}$$

For the exit task t_{exit}, its *sblevel* is

$$sblevel(t_{exit}) = w(t_{exit})/\overline{w(p)}. \tag{4.11}$$

Algorithm 2 is used to calculate *sblevel* for all tasks in the workflow parallel application DAG.

As an illustration, Table 4.4 outlines the *sblevel* values produced by Algorithm 2 for the sample worklfow parallel application DAG in Figure 4.1 on heterogeneous computing systems with processor average computation capacity $\overline{w(p)} = 1$ [81].

TABLE 4.4 The *sblevel* of sample DAG tasks in Figure 4.1

t_1	t_2	t_3	t_4
$N(24.20, 26.80)$	$N(17.76, 33.20)$	$N(17.86, 18.00)$	$N(11.10, 29.00)$
t_5	t_6	t_7	
$N(8.00, 14.00)$	$N(13.00, 7.25)$	$N(4.00, 4.00)$	

Input: A tasks set of workflow application DAG.
Output: The tasks *sblevel*.
1 Construct a list of tasks in a reversed topological order and call it
  RevTopList;
2 Compute *sblevel*(t_{exit}) using Eq. (4.11) and remove task t_{exit} from
  RevTopList;
3 **while** *the RevTopList is not empty* **do**
4   Remove the first task t_x from *RevTopList*;
5   Get a child t_i of t_x and use Eq. (4.5) to compute the expected value and
        variance of *sblevel*(t_x)←$w(a_{x,i})$+*sblevel*(t_i);
6   **for** *each other child t_i of t_x* **do**
7   Calculate the expected value and variance of *sblevel*(t_x) using Eqs.
          (4.7) and (4.8);
8   **end**
9   Construct an approximate normal distribution of *sblevel*(t_x);
10   Compute *sblevel*(t_x) using Eq. (4.5);
11 **end**

Algorithm 2: Computing the DAG tasks *sblevel*

4.4.2   Stochastic Dynamic Level Scheduling Algorithm

This section describes the Stochastic Dynamic Level Scheduling (SDLS) algorithm, which is derived from traditional deterministic DLS [75, 120]. This algorithm is based on *stochastic dynamic level* (*SL*) that to be defined as follows [81].

For heterogeneous cluster computing systems, the processors computation capacity are different from each other. Therefore, it is defined as

$$\Psi(t_i, p_k) = w(t_i)/\overline{w(p)} - w(t_i)/w(p_k). \qquad (4.12)$$

A large positive $\Psi(t_i, p_k)$ denotes that processor p_k is faster than most processors of systems, while a large negative $\Psi(t_i, p_k)$ expresses the opposite case [81]. The $SL(t_i, p_k)$ of task t_i on processor p_k is expressed as task stochastic bottom level *sblevel*(t_i) subtracted by task earliest execution start time $EES(t_i, p_k)$ on processor p_k, and added by $\Psi(t_i, p_k)$ [81], as given by the following equation,

$$SL(t_i, p_k) = sblevel(t_i) - EES(t_i, p_k) + \Psi(t_i, p_k). \qquad (4.13)$$

Input: The workflow application DAG tasks.

Output: A schedule solution s.

1 Calculate *sblevel* of each task using Algorithm 2;

2 Push the entry task t_{entry} into the ready task pool;

3 **while** *the ready task pool is not empty* **do**

4   **for** *each task t_i in the ready task pool* **do**

5   **for** *each processor p_k* **do**

6   Compute $SL(t_i, p_k)$ using Eq. (4.13);

7   **end**

8   **end**

9   Find the optimal task-processor pair (t_i, p_k) with high $SL(t_i, p_k)$;

10   Remove task t_i from the ready task pool;

11   Assign task t_i to processor p_k;

12   Push the unconstrained child tasks of t_i into the ready task pool;

13   Update the tasks earliest execution start time on processors;

14 **end**

Algorithm 3: The stochastic dynamic level scheduling algorithm

The SDLS is formally described in Algorithm 3. Similar to the traditional deterministic DLS algorithm, in each step of the mainly while loop, the SDLS algorithm first uses Eq. (4.13) to compute the all ready tasks SL on all possible processors in cluster computing systems [81]. Then, the optimal task-processor pair (t_i, p_k), where the maximum $SL(t_i, p_k)$ is selected and assign task t_i to the corresponding processor p_k. The main difference between stochastic SDLS and traditional DLS is how to find the optimal task-processor pair with the maximum SL, because the key attribute SL are random variables with normal distributions in this chapter [81].

Here, we define an operator $\Re$ to obtain the maximum SL for task-processor pair. We let function $F_Y(y)$ indicate the cumulative density of a random variable Y [81]. We assume that Y_1 and Y_2 are two random variables with normal distributions, and y_1 is the point such that $F_{Y_1}(y_1) = 0.9$, and y_2 is the point such that $F_{Y_2}(y_2) = 0.9$. We say that Y_1 is *stochastically less than* Y_2, or Y_2 is *stochastically greater than* Y_1, denoted by $Y_1 \Re Y_2$, if and only if $y_1 < y_2$. The SDLS algorithm finds the optimal task-processor pair (t_i, p_k) whose $SL(t_i, p_k)$ is greater than others by operator $\Re$. Figure 4.3 shows two cumulative density functions, which the variable Y_2 is stochastically greater than Y_1. We think that the probability 0.9 is high enough to yield good scheduling solution for SDLS algorithm [81].

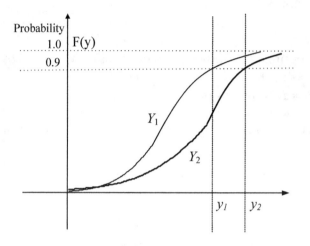

Figure 4.3   The operator $\Re$.

4.4.3   An Illustration Example

To illustrate the effectiveness of SDLS algorithm, we consider the stochastic tasks in the workflow application DAG shown in Figure 4.1 and a cluster computing systems with three processors p_0, p_1, and p_2, where $w(p_0) = 0.7$, $w(p_1) = 1.4$, and $w(p_2) = 0.9$ [81]. The schedule solution s produced by the DLS algorithm [120], [75], which is based on tasks mean execution and data communication time, and the algorithm SDLS, are shown in Figures 4.4(a) and 4.4(b), respectively. The task execution time and data communication time of Figure 4.4 are obtained from their actual values on cluster systems. For example, the execution time of task t_1 on processor p_1 is 2.27, and the execution time of task t_4 on processor p_2 execution time is 5.45 [81]. The solution produced by the SDLS algorithm has schedule length 14.46, which is shorter than that of DLS algorithm (15.33).

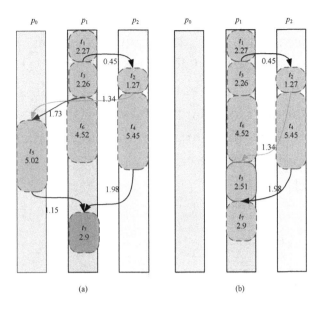

Figure 4.4 An illustration example. (a) DLS (makespan = 15.33); (b) SDLS (makespan = 14.46).

4.5 WORKFLOW APPLICATION PERFORMANCE EVALUATION

In this section, we compare the performance of SDLS with two existing scheduling algorithms: Rob-HEFT [23] and HEFT [146]. In order to make the comparison fair, we slightly modify the deterministic HEFT algorithm to take the expected value of task execution and edge data communication time as parameters [81]. We conduct a simulation environment for heterogeneous cluster computing systems with 16 processors, whose computation capacities are in the range from 2,000 MIPS to 3,000 MIPS. The comparison of the algorithms are based on the two performance metrics: the makespan (or schedule length), which is defined as the finish time of the *exit* task, and the speedup (see Eq. (3.8) in Chapter 3) [81].

4.5.1 Special Application DAG

In this simulation experiments, we examine the scheduling performance of SDLS with Rob-HEFT, and HEFT algorithms by some special parallel application workflow DAGs. One special DAG in

Figure 4.5(a) has 50 tasks of width no more than 4 nodes, which represents a parallel application with very low parallelism degree [81]. The other special DAG in Figure 4.5(b) has 50 tasks with 3 levels, which represents a parallel application with very high parallelism degree. The stochastic tasks execution size and data communication are varying parameters such as the minimum and maximum expected values (μ_{min}, μ_{max}) and variances $(\sigma_{min}, \sigma_{max})$ [81].

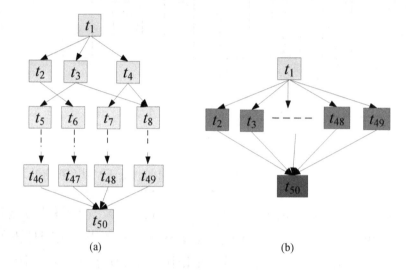

(a) (b)

Figure 4.5 Examples of workflow DAGs. (a) A low parallelism degree application; (b) A high parallelism degree application.

4.5.2 Experimental Results

Tables 4.5, 4.6, 4.7, and 4.8 show the simulation results of Figure 4.5 on heterogeneous cluster systems. From Table 4.5, we observe that the three scheduling algorithms have almost the same performance for low parallelism degree applications. As the number of processors increases, such as 10, 12, 14, 16, the makespan of these algorithms are almost the same [81]. Similarly, the experimental results (Table 4.6) about speedup with Figure 4.5(a) has the similar conclusions. However, from Tables 4.7 and 4.8, we observe that SDLS algorithm significantly outperforms Rob-HEFT by 7.7%,

HEFT by 7%, respectively, in term of the average makespan. For the average speedup, SDLS outperforms Rob-HEFT by 7.3% and HEFT by 6.9%.

TABLE 4.5 The special workflow DAG experimental results about makespan with Figure 4.5(a)

Processor Number	4	6	8	10	12	14	16
SDLS	194.4	227.4	227.4	206.2	206.2	206.2	206.2
Rob-HEFT	201.5	227.4	227.4	206.2	206.2	206.2	206.2
HEFT	194.4	227.4	227.4	213.4	213.4	206.2	206.2

TABLE 4.6 The special workflow DAG experimental results about speedup with Figure 4.5(a)

Processor Number	4	6	8	10	12	14	16
SDLS	2.52	2.16	2.16	2.39	2.39	2.39	2.39
Rob-HEFT	2.44	2.16	2.16	2.39	2.39	2.39	2.39
HEFT	2.52	2.16	2.16	2.3	2.3	2.39	2.39

TABLE 4.7 The special workflow DAG experimental results about makespan with Figure 4.5(b)

Processor Number	4	6	8	10	12	14	16
SDLS	167.5	134.8	110.3	103.2	95.4	93.1	90.6
Rob-HEFT	184.3	147.2	127.3	108.4	100.6	97.25	91.3
HEFT	179.65	149.6	126.4	109.6	99.8	95.1	90.6

This is due to the fact that the SDLS algorithm considers the expected value and variance of workflow application task execution and edge data communication time, and produces better schedule solution for the stochastic scheduling problem [81]. However, the other scheduling strategies do not comprehensively consider the random dispersion of the workflow parallel application task execution and edge data communication time, so we are

TABLE 4.8   The special workflow DAG experimental results about speedup with Figure 4.5(b)

Processor Number	4	6	8	10	12	14	16
SDLS	2.98	3.54	4.39	4.69	5.23	5.37	5.63
Rob-HEFT	2.5	3.31	3.77	4.55	5	5.19	5.321
HEFT	2.72	3.27	3.87	4.47	4.9	5.14	5.4

not suitable for stochastic scheduling on cluster computing systems. Finally, we point out that with the increase of the number of processors, the schedule length of the three scheduling algorithms decreases, and the speedup of them increase [81]. These results also show that the stochastic dynamic level scheduling algorithm is very suitable for high parallelism degree workflow applications.

4.6   SUMMARY

This chapter focuses on the importance of scheduling stochastic workflow parallel applications to the cluster computing systems. We presents an efficient scheduling algorithm, which employs stochastic dynamic level to produce schedule solution with high quality, to meet the increasing challenge from real-world applications with random task execution and data communication time. We also point out that the integration of workflow application stochastic is the key characteristic which affects the performance of task scheduling algorithm.

CHAPTER 5

Reliability-Energy-Aware Scheduling algorithm

IN recent years, the amount of energy required to run high-performance computing systems has been growing rapidly, and this type of energy consumption phenomena has also attracted a great deal of attention from academia and industry. Moreover, high energy consumption inevitably results in failures and reduces reliability for systems with massive computing resources. Therefore, it is necessary to comprehensively consider the optimal management of performance, reliability, and energy consumption on large-scale heterogeneous systems. In this chapter, we first describes the scheduling architecture, heterogeneous computing systems, workflow applications, and energy consumption model. Then, we present the optimal management problem of performance, reliability, computing processor frequencies, and energy consumption. Thirdly, a heuristic reliability-energy aware scheduling algorithm (REAS) is presented to deal with this problem.

5.1 INTRODUCTION

Over the past few decades, powerful computing systems have been developed to achieve higher performance due to the rapid growing IT demands from industry and academia. However, the

DOI: 10.1201/b23006-5

power consumption of such systems has increased dramatically and causes severe ecological, economic, and technical issues. According to the latest world's Top 500 supercomputers Ranking, the world's first rank supercomputer Fugaku on the Top 500 list has a maximum operating power consumption of 29.899MW [1]. Thus, it is obvious that high energy cost is a key feature of designing and applying heterogeneous systems. Therefore, high energy consumption is a key feature in the design and application of large-scale computing systems [142].

In addition, modern high-performance computing systems are usually supported by a set of processors, which are connected through high-speed interconnection networks to support the execution of parallel applications [142]. For instance, the second supercomputer Summit in the world Top 500 large computing systems lists consists of IBM POWER9 22C 3.07GHz and NVIDIA VOLTA GV100 [1]. Here, the number of transistors integrated into modern high-performance processor reaches to almost 2.3 billion and power consumption of such is over $130W$ [116]. The reliability of such high integration processors may be reduced, and eventually resulting in the poor reliability of the whole heterogeneous systems. In addition, the high-performance large-scale computing systems usually has a great deal of processors, such as, Fugaku with $7,630,848$ cores and Summit with $2,414,592$ cores [1]. The system reliability will drastically decreases as the number of whole cores increases [93]. Even if the one hour reliability of a single processor becomes very high, such as 0.999999, the MTTF (mean time to failure) of the system will drop below 10 hours when the system size is close to $10,000$ cores [93]. Noting this, this chapter focuses primarily on the main problem of simultaneous management of reliability, energy consumption, and system performance [142].

5.2 SYSTEM MODELS

This section builds the scheduling architecture, heterogeneous computing systems, workflow applications, and energy consumption model.

5.2.1 Task Scheduling Architecture

There are many workflow task scheduling architectures proposed in Refs. [82, 104, 134, 135, 139, 166]. All of them are not effectively incorporating the system reliability and energy consumption into task scheduling. We present a reliability-energy aware workflow scheduling architecture, which is depicted in Figure 5.1. Here, we assume that the user parallel applications and their configuration information are submitted to the system through a special user command. These parallel applications are first processed into workflow DAG data structure by Task DAG Model. The Energy Consumption Estimator is used to estimate each task's energy consumption, which executes on the DVFS-enabled processors [142]. At the same time, the Reliability Analysis will compute system processor's reliability according to its frequency and obtain the whole system reliability. The Scheduler schedules DAG tasks according to the above system reliability estimation and task energy consumption [142].

5.2.2 Heterogeneous Computing Systems

In this chapter, the target computing systems consist of a set $P = \langle p_1, p_2, \cdots, p_m \rangle$ heterogeneous DVFS-enabled machines or processors that connected by high-speed interconnection network, such as Myrinet and InfiniBand [82, 134, 135, 139, 166]. In this model, DVFS-enabled processor $p_k \in P$ can dynamically adjust its operational frequency and voltage [82]. Thus, the parallel applications task can be executed on discrete frequency-voltage pairs, $(f_{k,h}, V_{k,h})$, where $(f_{k,1} < f_{k,2} < \cdots < f_{k,M_k})$ and $(V_{k,1}, V_{k,2} < \cdots < V_{k,M_k})$. Here, the M_k is the number

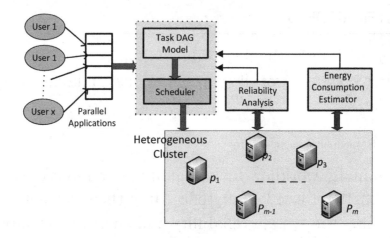

Figure 5.1 The reliability-energy aware workflow scheduling architecture.

of processor p_k's operation frequencies [82, 131]. For instance, the quad-core AMD Phenom II has four operation frequencies (3.2GHz, 2.5GHz, 2.1GHz, and 0.8GHz), and corresponding voltage is ranging from $1.425V$ to $0.85V$ [131].

In this model, the processor's failure is also assumed to follow the Poisson process, and processor is assigned a failure rate λ used in many work [134, 135, 142, 166]. Here, we let λ_k denotes processor p_k failure rate when it works at normal frequency and voltage. [40, 134, 135, 166]. These failure rate parameters can be derived from computing systems' log, profiling, and some statistical prediction techniques [105, 142]. For illustration purposes, we list two heterogeneous processors for example, one has three frequencies and the other has two frequencies [142], and these parameters are listed in Table 5.1.

TABLE 5.1 The parameters of computing systems processors

	λ_k	$\Phi_s[k]$	$\alpha[k]$	$(f_{k,h}, V_{k,h})$		
				1	2	3
p_1	1.4×10^{-4}	73.6	3.663×10^{-8}	$(0.8 \times 10^9,$ $0.93)$	$(2.1 \times 10^9,$ $1.23)$	$(3.2 \times 10^9,$ $1.43)$
p_2	1.62×10^{-4}	57.1	4.95×10^{-8}	$(2.3 \times 10^9,$ $0.85)$	$(3.0 \times 10^9,$ $1.36)$	

5.2.3 Parallel Application Workflow DAG

Generally, the workflow parallel application is expressed as a Directed Acyclic Graph (DAG) $G = \langle T, A, [a_{i,j}], [w_{i,k,h}] \rangle$ [134, 135, 139, 142, 146, 166], where $T = \{t_1, t_2, \cdots, t_n\}$ is a finite n tasks set that can be assigned to any available computing systems processors [134, 135, 139, 146, 166]; A indicates the partial priority of tasks, such that $t_i \, R \, t_j$ implies task v_i must execute finish before t_j can start its execution. $[a_{i,j}]$ is the data communication matrix and each denotes the time between task t_i and t_j for $1 \leq i, j \leq n$. $[w_{i,k,h}]$ is a $n \times m \times M_{max}$ computation matrix in which each $w_{i,k,h}$ gives the execution estimation time of task t_i on processor p_k at frequency $f_{k,h}$ [142]. Here, M_{max} denotes the maximal number of processor operation frequency for the whole systems. The above estimation time can be obtained by code profiling, historical table, and statistical prediction techniques [105]. This chapter shows an example of workflow application DAG in Figure 5.2, The tasks estimation execution time on two DVFS-enabled processors (see Table 5.1) are listed in Table 5.2. We also list the data communication time among these tasks in Table 5.3 [142].

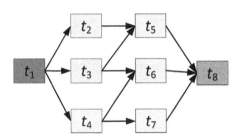

Figure 5.2 The example of workflow application DAG graph.

In general, the common goal of computing systems workflow scheduling is to assign precedence constrained tasks to processors and achieve the minimum makespan [142, 146, 165]. Before deriving the scheduling objective schedule length, we need to formally define some important scheduling attributes of task t_i.

TABLE 5.2 The task estimation execution matrix $[w_{i,k,h}]$

| Task | p_1 | | | p_2 | |
	$p_{1,1}$	$p_{1,2}$	$p_{1,3}$	$p_{2,1}$	$p_{2,2}$
t_1	11.12	2.28	2.87	3.89	3
t_2	36.29	13.82	9.12	12.7	9.78
t_3	15.46	5.91	3.92	5.45	4.21
t_4	5.33	2.01	1.4	1.94	1.49
t_5	66.77	25.44	16.77	23.28	17.83
t_6	13.82	5.3	3.53	4.84	3.75
t_7	7.43	2.86	1.89	2.68	2.04
t_8	8.48	3.19	2.09	2.91	2.31

Here, we let $EES(t_i, f_{k,h})$ indicates task t_i earliest execution starting time on processor p_k at frequency $f_{k,h}$. Generally, this time is affected by workflow tasks priority and the available time of processor [134, 135, 166]. Accordingly, $EEF(t_i, f_{k,h})$ is the earliest execution finish time, which is defined by

$$EEF(t_i, f_{k,h}) = EES(t_i, f_{k,h}) + w_{i,k,h}. \qquad (5.1)$$

TABLE 5.3 The estimation data communication matrix $[a_{i,j}]$

Task	t_2	t_3	t_4	t_5	t_6	t_7	t_8
t_1	6.99	15.48	6.69				
t_2				10.86			
t_3				1.25	12.56		
t_4					6.93	0.3	
t_5							0.11
t_6							6.535
t_7							6.2

This chapter lets $X_{i,k}^h = 1$ represents task v_i is scheduled on processor p_k at frequency $f_{k,h}$, otherwise $X_{i,k}^h = 0$. Therefore, the schedule length or makespan is described as follows,

$$makespan = MAX_{1 \le i \le n, 1 \le k \le m}^{1 \le h \le M_k}\{X_{i,k}^h EEF(t_i, f_{k,h})\}. \qquad (5.2)$$

5.2.4 Energy Consumption Model

For large-scale computing systems, the energy consumption depends on its computing resources (such as CPUs, GPUs, and so on), disks, memory, and other systems components [142]. This chapter only considers the maximum energy consumption of computing resources on the system. [56, 82, 110]. For modern DVFS-enabled microprocessor based on Complementary Metal Oxide Semiconductor (CMOS) logic circuits, the main power consumption consists of dynamic power dissipation and static power [178, 179], which is usually modeled as

$$\Phi = \Phi_s + \zeta\Phi_d, \tag{5.3}$$

where Φ_s denotes the static power that is a constant, and this power consumption is used to keep the clock running, maintain basic circuits, and frequency-independent active power. Here, ζ is the processor work model. If processor is work at execution model, $\zeta = 1$. Otherwise, $\zeta = 0$. In this model, Φ_d is the execution power consumption and consume the most significant factor of processor [56, 61, 82, 110], which can be approximately estimated by

$$\Phi_d = \kappa V^2 f. \tag{5.4}$$

Here, κ indicates the switched capacitance, f denotes the processor working frequency, and V is the supply voltage. The illustration parameters are listed in Table 5.1.

We also let $ENC(t_i, f_{k,h})$ denotes the energy consumption that consume by task t_i executing on processor p_k at frequency $f_{k,h}$. This energy is determined by processor power consumption and task execution time. We define it as

$$\begin{aligned} ENC(t_i, f_{k,h}) &= w_{i,k,h} \times \Phi^k \\ &= w_{i,k,h} \times \Phi_s^k + w_{i,k,h} \times \Phi_d(f_{k,h}), \end{aligned} \tag{5.5}$$

where $\Phi_d(f_{k,h})$ indicates dynamic power dissipation of processor p_k at frequency $f_{k,h}$ (see Eq. (5.4)). Therefore, for workflow parallel applications, the energy consumption $ENC(T)$ is the sum of the energy consumption of all tasks,

$$
\begin{aligned}
ENC(T) &= \sum_{\substack{1 \leq i \leq n, 1 \leq k \leq m}}^{1 \leq h \leq M_k} \{X_{i,k}^h ENC(t_i, f_{k,h})\} \\
&= \sum_{\substack{1 \leq i \leq n, 1 \leq k \leq m}}^{1 \leq h \leq M_k} \{X_{i,k}^h w_{i,k,h} \times \Phi_s^k + X_{i,k}^h w_{i,k,h} \times \Phi_d(f_{k,h})\}.
\end{aligned}
\tag{5.6}
$$

At the same time, for large-scale computing systems, whether processors are sleep or execution models, the systems always need to consume the static power energy consumption. Therefore, the systems energy consumption $ENC(P)$ is the summation of all processors dynamic power dissipation and static power of all tasks energy consumption,

$$
\begin{aligned}
ENC(P) =& makespan \times \sum_{k=1,2,\cdots,m} \Phi_s^k + \\
& \sum_{\substack{1 \leq i \leq n, 1 \leq k \leq m}}^{1 \leq h \leq M_k} \{X_{i,k}^h w_{i,k,h} \times \Phi_d(f_{k,h})\}.
\end{aligned}
\tag{5.7}
$$

Evidently, the application energy consumption $ENC(T)$ is less than systems energy consumption $ENC(P)$. One of mainly goals is to minimize systems energy consumption $ENC(P)$.

5.3 SYSTEM RELIABILITY ANALYSIS

This section deduces the single processor failure rate model and analyzes the high-performance systems reliability.

5.3.1 Single Processor Failure Rate

Among various unreliable sources of semiconductor circuit processor, it is predicted that the failure rate due to cosmic ray radiation soft errors will be the main factor in reliability [52]. Semiconductor device transient fault will occur when the high energy particle, such as neutron or alpha, strikes a sensitive region and flips the logical state of the struck node [15,142]. Most of the modern semiconductor processor is the integration of multi-billion transistors on a single chip. In such case, it results in the increasing number of sensitive devices that is vulnerable to soft error and consequently raises the *Soft Error Rate* (SER). With the continuous change of processor frequency and voltage, this phenomenon becomes more and more serious [50, 178].

When the semiconductor processor works at normal frequency and voltage, its reliability is usually modeled as following the Poisson distribution with a failure rate λ [40, 130, 135, 142, 166]. In addition, it has been fully shown that blindly applying DVFS technique to adjust power supply for energy savings has a negative impact on processor failure rate, which may lead to the significant degradation in processor reliability [50, 178, 179]. Thus, for the semiconductor processor $p_k \in P$ considering in this chapter, the failure rate at a reduced frequency $f_{k,h}$ is defined as follows,

$$\lambda_k(f_{k,h}) = \lambda_k \cdot \Upsilon_k(f_{k,h}). \tag{5.8}$$

Here, λ_k is the failure rate corresponding to the normal processing frequency f_{norm}. Previous study works on the impact of frequency on processor reliability have shown that the failure rate usually increases with the scaling of processing frequency away from the normal [36, 52, 142]. On the other hand, the failure rate is exponentially related to the critical charge (i.e. threshold voltage) of circuit [142]. Therefore, we have the following equations,

$$\Upsilon_k(f_{k,h}) = \begin{cases} e^{\varpi_k V t_k} 10^{\varrho_k \frac{f_{k,h} - f_{norm}}{f_{max} - f_{min}}} & f_{norm} \leq f_{k,h} \leq f_{max}, \\ e^{\varpi_k V t_k} 10^{\varrho_k \frac{f_{norm} - f_{k,h}}{f_{max} - f_{min}}} & f_{min} \leq f_{k,h} \leq f_{norm}. \end{cases} \tag{5.9}$$

Here, the ϱ_k is a constant and exponent ϖ_k is the parameter of threshold voltage, where they are denoted as the sensitivity of fault rates by frequency scaling. Furthermore, f_{max} and f_{min} indicate the maximum and minimum frequencies, respectively.

This chapter uses least squares curve fitting method to obtain the precision value of parameters ϱ_k and ϖ_k [79, 142]. First, the natural logarithm of both sides for Eq. (5.9) is

$$ln(\Upsilon_k(f_{k,h})) = \begin{cases} \varpi_k V t_k + \varrho_k \ln 10 \frac{f_{k,h} - f_{norm}}{f_{max} - f_{min}} & f_{norm} \leq f_{k,h} \leq f_{max}, \\ \varpi_k V t_k + \varrho_k \ln 10 \frac{f_{norm} - f_{k,h}}{f_{max} - f_{min}} & f_{min} \leq f_{k,h} \leq f_{norm}. \end{cases}$$

$$(5.10)$$

Here, we let $R = \varpi_k V t_k$, $U = \varrho_k \ln 10 \frac{1}{f_{max} - f_{min}}$, $y = \ln(\Upsilon_k(f_{k,h}))$, and $Z = \varrho_k \ln 10 \frac{f_{norm}}{f_{max} - f_{min}}$ [142]. Therefore, the Eq. (5.10) become

$$y = \begin{cases} R + U f_{k,h} - Z & f_{norm} \leq f_{k,h} \leq f_{max}, \\ R - U f_{k,h} + Z & f_{min} \leq f_{k,h} \leq f_{norm}. \end{cases}$$

$$(5.11)$$

In this way, we can obtain the approximation value of parameters ϱ_k and ϖ_k by using least squares linear fitting approach.

5.3.2 Application Reliability Analysis

This chapter assumes that the execution time of task t is taken place during the time interval $[B, C]$ on processor p_k at frequency $f_{k,h}$ [142]. Here, B represents task start execution time and C is the task finish time [134, 135, 139, 166]. Therefore, the execution reliability of task t can be expressed as

$$\begin{aligned} P[t] &= P\{X(C) - X(B) = 0\} \\ &= P\{X(C - B + B) - X(B) = 0\} \\ &= exp\{-\lambda_k(f_{k,h})(C - B)\}. \end{aligned}$$
$$(5.12)$$

For workflow parallel application task t_i on processor p_k at frequency $f_{k,h}$, its execution reliability $P[t_i, f_{k,h}]$ is equal to all of its immediate parent tasks and itself execution reliability [142], which can be given by,

$$P[t_i, f_{k,h}] = \prod_{t_j \in prev(t_i)} P[t_j] \times exp(-\lambda_k(f_{k,h}) \times w_{i,k,h}),$$
$$(5.13)$$

where $prev(t_i)$ represents all direct predecessors of t_i, $P[t_j]$ is the reliability of task v_j that is equal to the reliability of task t_i executing on processor p_k at frequency $f_{k,h}$,

$$P[t_j] = \sum_{1 \leq k \leq m}^{1 \leq h \leq M_k} \{X_{1,k}^h P[t_j, f_{k,h}]\}.$$
$$(5.14)$$

For the entry task t_1 executing on processor p_k at frequency $f_{k,h}$, its $prev(t_1) = \phi$, and the reliability is

$$P[t_1, f_{k,h}] = exp(-\sum_{\substack{1 \le k \le m \\ 1 \le h \le M_k}} \{X_{1,k}^h \lambda_k(f_{k,h}) \times w_{1,k,h}\}). \qquad (5.15)$$

Generally, workflow application has one exit task t_{exit} [142]. Thus, the reliability of application $P[G]$ is equal to the exit task t_{exit},

$$\begin{aligned} P[G] &= P[t_{exit}] \\ &= \prod_{t_j \in prev(t_{exit})} P[t_j] \times P[t_{exit}, f_{k,h}]. \end{aligned} \qquad (5.16)$$

This is another goal of this chapter. We try to improve the reliability of applications reliability $P[G]$. From the above analysis, we know that assigning tasks with shorter execution time to more reliable processors may be a good way to improve reliability [142].

As simultaneous management of system reliability, energy consumption, and scheduling performance is the main problem of this chapter [142], and we formulate it as

$$\begin{cases} Minimize\ makespan, \\ Minimize\ ENC(P), \\ Maximize\ P[G], \\ s.t. \\ X_{i,k}^h = 1\ Or\ X_{i,k}^h = 0 \\ \sum_{\substack{1 \le k \le m \\ 1 \le h \le M_k}} X_{i,k}^h = 1 \quad \forall\ t_i \in T \\ t_i\ A\ t_j \quad \forall\ t_i, t_j \in T \end{cases} \qquad (5.17)$$

5.4   THE RELIABILITY-ENERGY AWARE SCHEDULING ALGORITHM

This section describes a reliability-energy aware scheduling algorithm (REAS) on high-performance large-scale computing systems, which aims at obtaining higher reliability, lower energy consumption, and shorter schedule length [142]. The workflow application task scheduling decision is made by using a hybrid metric integrating reliability, energy consumption, and makespan. The pseudo-code of the REAS algorithm is shown in Algorithm 4.

5.4.1 Task Priorities Phase

The tasks priority calculation is an important phase of list scheduling algorithms. The task list is usually generated by sorting tasks in decreasing order of some predefined rank functions, such as *b_level, t_level, CP, Rank,* and *DL* [43, 134, 135, 142, 146]. This chapter adopts *b_level* as its rank function. Primitively, the task *b_level* is the sum of the path weight from current task to exit task. It can recursively traverse the DAG graph from the exit task to calculate this value, and which is defined by

$$b\_level(t_i) = \overline{w(t_i)} + MAX_{t_j \in next(t_i)}\{a_{i,j} \\ + b\_levle(t_j)\} + RO(t_i), \tag{5.18}$$

where the $next(t_i)$ denotes the set of immediate successors of task t_i. In this priority calculation, we define the average computation capacity on task t_i as $\overline{w(t_i)}$, which is expressed as

$$\overline{w(t_i)} = \frac{\sum_{1 \leq k \leq m}^{1 \leq h \leq M_k} w_{i,k,h}}{\sum_{1 \leq k \leq m} M_k}. \tag{5.19}$$

Here, the $RO(t_i)$ is the average reliability overhead of task t_i and can be defined by,

$$RO(t_i) = \left(1 - exp\left\{-\frac{\sum_{1 \leq k \leq m}^{1 \leq h \leq M_k} \lambda_k(f_{k,h})}{\sum_{1 \leq k \leq m} M_k} \times \overline{w(t_i)}\right\}\right) \times \overline{w(t_i)}. \tag{5.20}$$

For the exit task t_{exit}, the *b_level* can be expressed as

$$b\_level(t_{exit}) = \overline{w(t_{exit})} + RO(t_{exit}). \tag{5.21}$$

In essence, the $b\_level(t_i)$ is the length of the critical path from current task t_i to the exit task, including task reliability overhead and average computation cost. For instance, considering the

Input: The DAG task set of workflow applications
Output: The task-processor pairs

1 Calculate each task b_level of DAG;
2 Sort tasks in a scheduling list by non-increasing order of b_level;
3 **while** *the scheduling list is not empty* **do**
4   Remove the first task t_i from the scheduling list;
5   Set $minF(t_i)$, $minE(t_i)$ as maximum value;
6   **for** *each processor-frequency $f_{k,h}$ in systems* **do**
7   Compute the earliest finish time $EEF(t_i, f_{k,h})$ using Eq. (5.22);
8   **if** $minRE(t_i) > EEF(t_i, f_{k,h})$ **then**
9             $minRE(t_i) \leftarrow EEF(t_i, f_{k,h})$;
10   **end**
11   Compute task energy consumption $EC(t_i, f_{k,h})$ using Eq. (5.5);
12   **if** $minRE(t_i) > EC(t_i, f_{k,h})$ **then**
13             $minRE(t_i) \leftarrow EC(t_i, f_{k,h})$;
14   **end**
15   **end**
16   Set $minRE(t_i)$ as maximum value;
17   **for** *each processor-frequency $f_{k,h}$ in systems* **do**
18   Compute the earliest finish time $EEF(t_i, f_{k,h})$ using Eq. (5.22);
19   Compute task energy consumption $EC(t_i, f_{k,h})$ using Eq. (5.5);
20   Compute hybrid metric $RE(t_i, f_{k,h})$ using Eq. (5.24);
21   **if** $minRE(t_i) > RE(t_i, f_{k,h})$ **then**
22             $minRE(t_i) \leftarrow RE(t_i, f_{k,h})$;
23   **end**
24   **end**
25   Assign task t_i to the corresponding processor-frequency;
26   Update the processor execution finish time;
27 **end**
28 **for** *each task in scheduling task-processor pairs* **do**
29   Compute task slack time $Slk(t_i)$ using Eq. (5.25);
30   **for** *each frequency of processor k* **do**
31   Compute the optimal frequency $f_{k,h}$ using Eq. (5.26);
32   **end**
33   Reassign task t_i and update corresponding information;
34 **end**
35 Compute systems energy consumption $ENC(P)$, schedule length, application reliability $P[G]$;

Algorithm 4: The reliability-energy aware scheduling algorithm

computing systems processor parameters in Table 5.1, workflow parallel application DAG in Figure 5.2, task execution time matrix in Table 5.2, and data communication matrix in Table 5.3. The tasks b_level value are recursive computed by Eq. (5.19), Eq. (5.21) that is shown in Table 5.4 [142].

TABLE 5.4   The workflow DAG task b_level value

Task	t_1	t_2	t_3	t_4	t_5	t_6	t_7	t_8
b_level	73.4	61.4	42.4	26	34.15	16.6	13.7	3.8
Seq	1	2	3	5	4	6	7	8

5.4.2   Task Assignment Phase

In this phase, tasks are scheduled to the computing systems processors with high reliability, minimum energy consumption $ENC(t_i)$, and earliest execution finish time $EEF(t_i)$. However, the above performance evaluation metrics are conflicted in most time [142]. Therefore, we introduce a hybrid scheduling objective named as RE, which can achieve good tradeoff among these performance metrics. Here, we redefine the task t_i earliest execution finish time on processor p_k at frequency $f_{k,h}$ as

$$EEF(t_i, f_{k,h}) = EES(t_i, f_{k,h}) + w_{i,k,h} + relO(t_i, f_{k,h}). \quad (5.22)$$

Here, the $relO(t_i, f_{k,h})$ is reliability overhead of task t_i on processor p_k at frequency $f_{k,h}$, and defined by

$$relO(t_i, f_{k,h}) = (1 - P[t_i, f_{k,h}]) \times w_{i,k,h}. \quad (5.23)$$

Furthermore, we let $minENC(t_i)$ and $minEEF(t_i)$ represent the minimum task energy consumption and earliest execution finish time on all systems processors, respectively [142]. Therefore, the hybrid task metric RE on processor p_k at frequency $f_{k,h}$ is

$$RE(t_i, f_{k,h}) = \theta \times \frac{EEF(t_i, f_{k,h}) - minEEF(t_i)}{EEF(t_i, f_{k,h})}$$
$$+ (1 - \theta) \times \frac{ENC(t_i, f_{k,h}) - minENC(t_i)}{ENC(t_i, f_{k,h})},$$
$$(5.24)$$

where, θ is the task weight to measure energy consumption and earliest execution finish time. Here, if the task energy consumption is more important than its execution time, it gives lower value to θ; otherwise, it sets θ with higher value. In addition, the scheduling goal of this study is try to obtain the minimum schedule length and energy consumption [142]. Therefore, we try to achieve the minimum $RE(t_i, f_{k,h})$ by assigning task to the corresponding processor-frequency in each step.

5.4.3 Slack Reclamation

The workflow parallel application tasks have precedence constraints that results in some periods of processor idle. For instance, multidimensional inter task communication (or inter task data dependency), and the idle time of these processors is an obvious source of energy wastage [142]. Slack reclamation techniques is a good method to reduce energy using the slack left by some completed task instances. The idea behind the slack reclamation for the saving of processor energy consumption is to exploit these slack time to slowdown the execution frequencies [78, 155]. This chapter adopts this technology to reduce energy consumption after the scheduling decision is made [142]. Here, the slack time of task t_i is computed by

$$Slk(t_i) = MIN_{t_j \in next(t_i)}$$
$$\begin{cases} Sch(t_j, sT) - Sch(t_i, fT), & if \ t_i, t_j \ on \ any \ processors \\ Sch(t_j, sT) - a_{i,j} - Sch(t_i, fT), & otherwise \end{cases}$$

$$(5.25)$$

where, the $Sch(t_i, sT)$ denotes the task t_i earliest start time on processor and $Sch(t_i, fT)$ is the earliest finish time [142]. If the task slack time is satisfied with $Slk(t_i) > 0$, it can scale down the processor execution frequency to save energy consumption [142]. Therefore, a better frequency $f_{k,h}$ must satisfy the following inequality,

$$\begin{cases} w_{i,k,h} + relO(t_i, f_{k,h}) < Sch(t_i, fT) + Slk(t_i), \\ ENC(t_i, f_{k,h}) < ENC(t_i, f_{k,orig}). \end{cases} \tag{5.26}$$

In this inequality, the $f_{k,orig}$ represents the original scheduling processor-frequency pairs [142]. Lastly, we reassign task t_i to the better frequency $f_{k,h}$.

5.5 EXPERIMENTAL RESULTS AND DISCUSSION

This section compares the energy consumption, scheduling performance, and system reliability using the reliability-energy aware scheduling algorithm with three existing scheduling algorithms: RDLS [40], ECS [78], and DLS [43]. The experiments are conducted by the synthetic randomly generated parallel application workflow DAG as describe in the below [142]. The performance evaluation metrics are the systems energy consumption $ENC(P)$ (Eq. (5.7)), schedule length (Eq. (5.2), Eq. (5.22)), and application reliability $P[G]$ (Eq. (5.16)) [142].

5.5.1 Simulation Environment

To test the above algorithms performance, we develop a simulator with 8 processors, which includes 2 Intel Xeon, 2 AMD Athlon, 2 Intel® Core^{TM} Duo, 1 Tesla GPU, and 1 TI DSP [142]. Other parameters of this simulation environment are set as follows: The failure rates of processors are assumed to be uniformly distributed between 1×10^{-4} and 1×10^{-5} failures/hr [104, 134, 135, 142], the interconnection link transmission rates are assumed to be 1000 Mbits/sec.

5.5.2 Randomly Generated Application

This experiments use three commonly DAG characteristics to generate parallel application graphs (some details of them can be seen in Section 3.5.2).

- **DAG size (v)**: The number of application DAG tasks [142].

- **Communication-Computation-Ratio** (CCR): It is the ratio of data communication time to task computation time. A small CCR value means the application is computation-intensive; a large CCR value denotes that the application is communication-intensive [134, 135, 142].

- **Out degree**: Out degree of DAG task node.

In the following experiments, the workflow DAG Graphs are randomly generated based on the above characteristics with the number of tasks 50 and 100 [142]. Task weights are generated from a uniform distribution $[1 \times 10^9, 9 \times 10^{11}]$ execution cycles to be approximately around 4.5×10^{10}. The DAG edge weights are also uniform distribution based on a mean CCR [142].

5.5.3 Various Weight θ of REAS Algorithm

We first evaluate the performance of weight θ to reliability-energy aware scheduling algorithm. Tables 5.5, 5.6, and 5.7 show the results of scheduling 50, 100 tasks with CCR=1. Here, the algorithms vary the weight θ from 0 to 1, in steps of 0.2. From Tables 5.5 and 5.6, we can conclude that the schedule length, energy consumption decrease as weight θ increases, and Table 5.7 shows that the reliability of application is almost at the same level [142]. It is due to the fact that the reliability-energy aware scheduling algorithm with high θ mostly based on task execution time, which results in shorter schedule length and less energy consumption [142]. However, if the weight θ is larger than 0.4, the performance of REAS has not much distinguishable. Therefore, in the following experiments, we let $\theta = 0.5$ [142].

TABLE 5.5 The schedule length of REAS algorithm with various weight θ

θ	0	0.2	0.4	0.6	0.8	1
50 Tasks	699.9	341.1	203.5	182.7	182.3	180
100 Tasks	2223.6	1412.6	518.5	512.4	502.9	459

TABLE 5.6 The energy consumption of REAS algorithm with various weight θ

θ	0	0.2	0.4	0.6	0.8	1
50 Tasks	388821.1	236652.9	203956	194914	222844.1	236906.7
100 Tasks	1231195.3	877462	596995.3	640740.4	688124.8	734847.6

TABLE 5.7 The reliability of REAS algorithm with various weight θ

θ	0	0.2	0.4	0.6	0.8	1
50 Tasks	0.978	0.978	0.977	0.974	0.97	0.969
100 Tasks	0.857	0.845	0.846	0.844	0.836	0.818

5.5.4 The Real-World Applications Results

It is very common to test the performance of algorithms using real applications [134,135,139,146,166]. The experimental results of digital signal processing (DSP) problem are shown in Figure 5.3. We observe Figure 5.3(a) that the REAS algorithm outperforms ECS and RDLS by 8.5%, 8.7% in term of schedule length, respectively [142]. In Figure 5.3(b), the REAS algorithm outperforms ECS, RDLS, and DLS by 4.3%, 6.3%, and 7.1% in term of

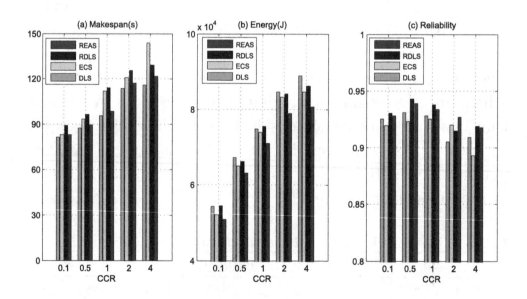

Figure 5.3 The experimental results of real-world DSP problem. (a) schedule length; (b) energy consumption; (c) reliability.

energy consumption, respectively [142]. In Figure 5.3(c), the four algorithms are also at the same level for reliability.

5.6 SUMMARY

In the recent years, with the rapid development of large-scale computing systems, the low reliability, high energy consumption, systems performance, and various environmental problems force the high-performance computing departments to reconsider some of their old practices in order to create more sustainable systems. We first present an energy-reliability aware workflow application task scheduling architecture, which mainly consists of systems, workflow DAG, and energy consumption model. Then, the relationship between reliability and processor frequency/voltage is presented. We also provide a heuristic reliability-energy aware scheduling algorithm to synthetic optimize energy consumption, time, and reliability.

CHAPTER 6

Energy Consumption and Reliability Bi-objective Workflow Scheduling

M̲ost studies on priority constrained workflow task scheduling in heterogeneous computing systems focus on low energy consumption and low execution time. In general, system reliability is more important than other performance metrics. In addition, power consumption and system reliability are two conflicting goals. How to compromise these two goals is a very important topic. This chapter introduces a new bi-objective genetic algorithm (BOGA) for workflow scheduling with low energy consumption and high system reliability.

6.1 INTRODUCTION

As introduced in Section 5.1, modern data centers consume a tremendous amount of power and produce a large amount of pollution. In addition, high energy consumption also has a negative effect on system reliability. High-performance heterogeneous computing system environments attract extensive attention of many researchers, who have conducted a large number of studies on diverse service quality of simple objectives based on heterogeneous computing environments; such objectives include low system delay, low energy consumption, minimum completion time, minimum execution cost, high system security, and high system reliability [154]. Such studies are more challenging when the focus is on more than one objective. Makespan and reliability bi-objective optimization can be transformed to a heuristic which targets makespan minimization while keeping reliability at a acceptable limit [44]. Boeres et al. developed an efficient weighted bi-objective scheduling algorithm for heterogeneous systems [20]. This chapter also introduced a classification of the solutions provided by weighted bi-objective schedulers aimed at helping users to adjust the weighting function such that an appropriate solution can be matched in accordance with different needs.

A large number of bi-objective algorithms are designed [10,41, 67,121,154]. Bi-objective optimization involves a combination of makespan and system reliability, execution time and cost, and makespan and energy consumption. The optimization of four objectives, including economic cost, makespan, energy consumption, and reliability, were preliminarily explored [51]. In the design process of multi-objective optimization algorithm, the idea of classical heuristic algorithm can be extended [154,171]. The multi-objective heterogeneous earliest finish time (MOHEFT [47]) list-based workflow scheduling algorithm aims at finding optimal solutions in terms of makespan and energy efficiency, which is an extension of the heterogeneous earliest finish time (HEFT [146]) algorithm. Meta-heuristic algorithms, such as artificial bee colony algorithms, are often used to expand the search space in multi-objective optimization.

However, none of the above studies consider low energy consumption and high system reliability at the same time. These two metrics are essential parts of modern green computing. Modern data centers need to provide a set of solutions with different performance metrics to meet the diverse demands of users, such as high system reliability and low energy consumption in different cases. Hence, the major purpose of this chapter is to present a general approach to solve the bi-objective of low energy consumption and high system reliability on workflow application scheduling in heterogeneous computing systems.

This chapter presents the bi-objective optimization problem of high system reliability and low energy consumption for parallel tasks as a combinatorial optimization problem. It includes three nontrivial subproblems, i.e., energy saving, reliability maximizing, and solution selecting. First, appropriate voltage levels are selected for processors so that the schedule length of parallel tasks is acceptable and the total energy consumption is minimal. Second, the selected speed for each parallel task is relatively high so that the total system reliability reaches a high level. Third, as energy saving and system reliability are two conflicting objectives, an effective and efficient solution strategy is devised while selecting the speed and processor for each candidate parallel task.

These problems are efficiently treated to develop an algorithm with an overall good performance. With the adoption of the non-dominated idea and DVFS technology, a novel bi-objective genetic algorithm that aims to achieve low energy consumption and high system reliability is developed [169].

6.2 MODELS AND PRELIMINARIES

In this section, the models used in this chapter are presented.

6.2.1 Workflow Model

A workflow application is usually modeled as a directed acyclic graph (DAG). A DAG, $W = (T, A)$ (the details can be seen in Section 1.3.2), where T is a finite set of tasks t_i $(1 \leq i \leq n)$ and $A = \{(t_i, t_j, Data_{ij}) | (t_i, t_j) \in T \times T\}$ is the set of edges which denote task precedence constrains, where $Data_{ij}$ is the volume of the data transferred from activity t_i to activity t_j. As shown in Figure 6.1, each directed arc is associated with a numeric label, which represents data flow.

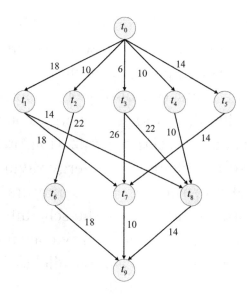

Figure 6.1 A simple DAG.

6.2.2 System Model

The system model consists of a set of *PE*, which includes p heterogeneous cores/processors in a cluster. Each processor $pe \in PE$ supports the DVFS technology wherein each processor works at several different speeds. For simplicity, the present chapter adopts the processor parameter configuration reported in [78], in which each *pe* has f different available frequency levels (AFLs). Since the frequency switching overhead occupies only an insignificant part of time, it is ignored in the following chapter. The voltage-frequency combinations used in this chapter are shown in Table 6.1. In addition, the communication subsystem explored in this chapter comprises a set of fully connected processors with each processor communicating with any other processors at any time that is completely free of contention.

6.2.3 Energy Model

The classic power model proposed in [178] is adopted to capture the system power in this chapter:

$$\Phi = \Phi_s + \zeta\left(\Phi_{ind} + \Phi_d\right) = \Phi_s + \zeta\left(\Phi_{ind} + \kappa f^\alpha\right), \qquad (6.1)$$

TABLE 6.1 Voltage-relative frequency combinations

Level	Combination 1		Combination 2		Combination 3	
	voltage	frequency	voltage	frequency	voltage	frequency
0	1.75	1.00	1.50	1.00	2.20	1.00
1	1.50	0.80	1.40	0.90	1.90	0.85
2	1.40	0.70	1.30	0.80	1.60	0.65
3	1.20	0.60	1.20	0.70	1.30	0.50
4	1.00	0.50	1.10	0.60	1.00	0.35
5	0.90	0.40	1.00	0.50	–	–
6	–	–	0.90	0.40	–	–

where Φ_s is the static power consumption, Φ_{ind} is the frequency-independent active power, and Φ_d is the frequency-dependent dynamic power. Static power includes the power to maintain basic circuits, to keep the clock working, and to keep memory to stay in sleep mode which can be removed only by turning off the whole system. Φ_{ind} is a constant, independent of system operation frequency (i.e., power consumption occurs while accessing external devices like the main memory, I/O, and so on), that is decreased to a very small value by setting the system to standby mode [22]. Φ_d is the dynamic power dissipation which is the dominant component of energy consumption in widely popular CMOS technology. It is shown by $\Phi_d = \kappa \cdot V^2 \cdot f$, where κ is the switched capacitance, V is the supply voltage, and f is the clock frequency. Since $f \propto V^\gamma$ $(0 < \gamma < 1)$ [83], in other words, $V \propto f^{1/\gamma}$, it is reckoned that the frequency-dependent active power consumption is $\Phi_d \propto f^\alpha$, where $\alpha = 1 + 2/\gamma \geq 3$. In the present chapter, $\Phi_d = \kappa f^\alpha$ is used. ζ denotes the system mode and represents the active power consumption that presently occurred. Particularly, $\zeta = 1$ indicates that the system is currently active. Otherwise, $\zeta = 0$ refers to a sleep mode that the system is in. In the context of this chapter, all frequencies are normalized with respect to the maximum frequency $f_{\max}$ (i.e., $f_{\max} = 1.0$).

The energy consumption of task t_i can be calculated by

$$E_i(f_i) = \Phi_{ind_i} \cdot \frac{w_i}{f_i} + \kappa \cdot w_i \cdot f_i^2, \qquad (6.2)$$

where w_i is the computation cost of the task t_i.

6.2.4 Reliability Model

As an application is running, a fault is hard to avoid due to hardware failure, software bugs, devices that work in high temperature, and so on. Accordingly, transient faults happen more frequently. Based on the previous study [171], the reliability model used in this chapter can be formalized as follows:

Definition 1 *The reliability of a task is the probability of executing the task successfully. If the transient fault follows a Poisson distribution, the reliability of node t_i with the corresponding computation cost w_i is [180]*

$$R_i(f_i) = e^{-\lambda(f_i) \times \frac{w_i}{f_i}}, \qquad (6.3)$$

where f_i denotes the processing frequency, $\lambda(f) = \lambda_0 \cdot g(f) = \lambda_0 \cdot 10^{\frac{d(1-f)}{1-f_{\min}}}$, d and λ_0 are constants.

6.2.5 Problem Definition

The problem addressed in the chapter consists of scheduling the workflow tasks applications in heterogeneous computing system under certain given constraints. The objectives are to minimize energy consumption and probability of failure (POF; equal to $1 - R$) while satisfying the shared deadline constraint of workflow tasks. This workflow scheduling problem is formalized as follows:

$$\text{Minimize:} \quad F = (F_1, F_2),$$

$$F_1 = \min_{t_i \in T,\, f_j(t_i) \in PE} Energy\,(t_i, f_j(t_i)),$$

$$F_2 = \min_{t_i \in T,\, f_j(t_i) \in PE} POF\,(t_i, f_j(t_i)),$$

$$\text{subject to:} \quad makespan(T) < D, \qquad (6.4)$$

where D is the shared deadline of the entire tasks set T.

To meet the task scheduling requirement, a prior order is established in this phase. Each task is set with its *URank*, which is computed recursively according to the following expression:

$$URank(t_i) = \overline{w_i} + \max_{t_j \in next(t_i)} \left(a_{i,j} + URank(t_j) \right), \qquad (6.5)$$

where $next(t_i)$ is the set of immediate successors of task node t_i, $a_{i,j}$ is the communication cost between nodes t_i and t_j and $\overline{w_i}$ is the average computational time of t_i when it is executed on different processors. Rank is recursively computed by traversing from the bottom to the top in a DAG. It should be apparent to draw such a conclusion that $URank(t_{exit}) = \overline{w_{exit}}$.

6.3 MULTI-OBJECTIVE OPTIMIZATION AND A MOTIVATIONAL EXAMPLE

In this section, several basic concepts of multi-objective optimization theory are introduced to understand this work better. Moreover, a motivational example is illustrated.

6.3.1 Multi-Objective Optimization Problem Overview

Since any maximization problem is transformed to a minimization problem, without loss of generality, the goal of all the objectives can be defined as minimization problem. Concepts from the multi-objective optimization (MOP) theory are introduced for a better understanding of this work. Since there are several performance metrics, such as energy saving, makespan, and system reliability in the workflow scheduling problem, it is usually formalized as a multi-objective optimization problem. In most cases, these metrics are a group of conflicting objectives when they are optimized at the same time. For Eq. (9.6), decreasing energy consumption brings an increment of the probability of failure. Classically, multi-object optimization problem is formally defined as follows:

Definition 2 *Given an n-dimensional vector $\vec{S} = [x_1, x_2, x_3, ..., x_n]$, MOP is to minimize the following objective function:*

$$\vec{F}(\vec{S}) = [f_1(\vec{S}), \ f_2(\vec{S}), \ ..., \ f_m(\vec{S})]. \qquad (6.6)$$

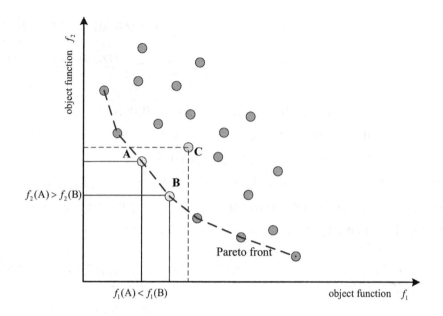

Figure 6.2 Multi-objective optimization.

Since finding a solution which minimizes both energy consumption and the probability of system reliability is not an easy work, the definition of dominance is introduced in the following.

Definition 3 *A solution is considered to dominate another solution if it is as good as the other and better in at least one objective. In other words, for two solutions $\vec{S}^*$ and $\vec{S}$ in the solution space, $\vec{S}^*$ dominates $\vec{S}$ if and only if Eq. (6.7) holds:*

$$\forall i \in [1,\ 2,\ 3,\ ...,\ m],\ f_i(\vec{S}^*) \leq f_i(\vec{S}) \wedge \exists j \in [1,\ 2,\ 3,\ ...,\ m],$$
$$f_j(\vec{S}^*) < f_j(\vec{S}).$$

$$(6.7)$$

As shown in Figure 6.2, solution A dominates solution C as the energy consumption and probability of system reliability of A are better (smaller) than those of C. Similarly, compared to solution C, solution B has smaller energy consumption and probability of system failure, so solution B dominates solution C. However, for solutions A and B, solution A is better in energy consumption and B is better in probability of system reliability. Hence, solutions A and B are non-dominated.

For a particular problem in this chapter, n denotes the number of the task set T ($|T| = n$), and the ith component of the solution $\vec{S}$ represents the resource on which task t_i is released for execution. In the bi-objective optimization problem scenario, $m = 2$ is obtained, where $f_1(\vec{S})$ is the energy consumption and $f_2(\vec{S})$ is the POF.

As reported in previous study [171], energy consumption and system reliability are two conflicting objectives. It is impossible to achieve the minimum for both POF and energy consumption simultaneously. In such problems, no single optimal solution exists but rather a set of potential solutions. A solution is considered to dominate another solution if it is as good as the other and better in at least one objective. Conversely, two solutions are considered to be non-dominated whenever neither of them dominates the other (one is better in energy saving and the other is better in system reliability).

Another important concept is the Pareto set which is a fine solution set. The Pareto set consists of a set of non-dominated solutions. It captures a set of tradeoff solutions among different objectives. Each solution in the Pareto set denotes a distinct task mapping of different processors with diverse energy consumption and POF. The Pareto front can be used as a tool to help the user make a decision on choosing the kind of mapping strategy for the workflow tasks resources.

6.3.2 A Motivational Example

When a DAG workflow model is submitted to a data center, the only constraint condition is the tasks completion time (makespan). As an illustration, the workflow is shown in Figure 6.1, and the computation cost of each task node on different processors is captured in Table 6.2. The constraint condition for this application is that the makespan does not exceed 135. While applying the BOGA, the data center provides the user with alternatives (shown by pairs in terms of energy

TABLE 6.2 Computation costs on different processors

Node number	P_1	P_2	P_3
0	11	16	10
1	12	17	11
2	10	12	15
3	12	6	11
4	14	13	17
5	13	8	12
6	9	12	11
7	14	10	17
8	15	17	10
9	10	18	16

consumption ratio and POF): (1.902854, 1.14e-06), (1.962252, 9.97e-07), (2.189743, 1.20e-06), (2.362726, 1.20e-06), (2.390827, 1.17e-06), (2.487445, 1.02e-06), (2.769279, 1.18e-06), (2.871348, 1.18e-06), and (3.051540, 1.24e-06). These solutions spread in the Pareto front.

6.4 ALGORITHMS

Several parts including encoding, initial population, fitness measure, selection, one-point crossover, mutation, and the main algorithm, respectively, are presented in this section. The novel algorithm, named BOGA, is devised in this section to address the workflow application scheduling problem in a heterogeneous cluster.

6.4.1 Encoding

In this approach, each chromosome comprises two components, which include the mapping and the scheduling string. The scheduling represents the DAG topological sort, which guarantees the precedence constrains. The chromosome length is equal to the number of the task set T ($|T| = n$). Let ch (a vector of length $|T|$) be the scheduling string. T_i appears only once in ch.

6.4.2 Initial Population

In the initial stage of BOGA, the quality of initial population is crucial. The strategy used in this stage is efficient and has a relatively low time complexity. Several outstanding heuristic list scheduling algorithms, i.e., HEFT [146], heterogeneous critical parents with fast duplicator (HCPFD) [59], predict earliest finish time (PEFT) [8] and so on, are proven to perform well in static task scheduling research. In the priority-establishing stage, these excellent strategies are selected randomly to generate distinct chromosomes.

6.4.3 Fitness Measure

A fitness function is used to measure the quality of the solutions according to the given optimization objectives. Two objectives which include system reliability and energy consumption are primary metrics in this chapter.

The energy consumption of a workflow application is measured by energy consumption ratio (ECR) expressed as follows:

$$ECR = \frac{E_{total}}{\sum_{t_i \in CP} \min_{pe_j \in PE}\left\{\Phi_{ind_i} \cdot \frac{w_i}{f_i} + \kappa \cdot w_i \cdot f_i^2\right\}}, \qquad (6.8)$$

where E_{total} is the total energy consumption of the scheduled tasks, and CP is the set of tasks in the DAG critical path.

A high system reliability problem can be transformed to a low POF. Hence, POF is used to measure the second objective of chapter. Apparently, POF is defined as:

$$POF = 1 - R = 1 - \Pi_{i=1}^{n} R_i(f_i). \qquad (6.9)$$

6.4.4 Selection

Applying the non-dominated sorting reported in [35], each solution is set with a *rank* of non-dominated solutions. According to this *rank*, all of the feasible solutions are classified. The non-dominated level of each candidate solutions is established. The solutions with equal *rank* values are placed in the same level. The smaller the *rank* value is, the higher is the solution priority which maintains elitism to the next iteration. For the particular objectives in this chapter, low POF and low energy consumption are the two main metrics. After the procedure of non-dominated sorting, *rank*1 is found to be at the inner position. From the inner to the outside positions, *rank*2, *rank*3, and so on followed. The solutions in the inner levels are prior to the one in the outside levels when the solutions are selected as an elite population for the next iteration.

To evaluate chromosome quality, i.e., the performance of both POF and energy consumption in the solution is very important. During algorithm iterations, two populations $(2n)$ are obtained after the mutation operation. The selection operation chooses the best n solutions from the candidate population (the size is $2n$). The rule to evaluate the solutions proceeds with the rank of solutions. The detailed rules of priorities selection, namely, the Selection Algorithm, are expressed as follows:

1. For solutions in different levels, the non-dominated solutions with lower level are preferred.

2. For the solutions in the same level, the solutions with large crowding distance are a priority than the small ones.

The selection procedure is shown in Figure 6.3.

6.4.5 Two-Point Crossover

The traditional crossover operation in genetic algorithm contains many kinds of strategies, such as one-point crossover, multi-point crossover, uniform crossover, cycle crossover, and so on. However, these common strategies are not directly applied to DAG chromosomes. The tasks sequence on a chromosome must satisfy the precedence constraints of the task set. After the completion of the cross operation, the tasks of the chromosome representation

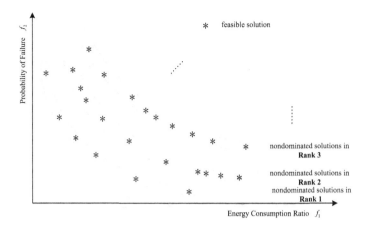

Figure 6.3 The selection procedure.

must maintain the topology of the task graph to ensure the correctness of the crossover strategy. For DAG, the following is a specific two-point cross way.

Let *parent*1 and *parent*2 be the two random chromosomes in the population, they are also the two feasible solutions to the DAG workflow task. Accordingly, the order of genes in each of the two chromosomes satisfies the precedence constraint of the task graph. *parent*1 and *parent*2 are the parent generation; after one-point crossover, two new chromosomes named *child*1 and *child*2 are generated. As shown in Figure 6.4, let T ($|T| = n$) be the length of the chromosome. The one-point crossover operation is expressed as follows:

1. Two positive integers are randomly picked up as the crossover points when the condition $0 \leq i \leq n - 1$ is met.

2. The genes in *parent*1 which locate before the crossover point i to *child*1 are located in the same location. When conditions $k < i$ and $0 \leq k \leq n - 1$ are satisfied, $child1_k = parent1_k$ is the result.

3. The genes in *parent2* which do not appear in child1 currently appeared and temporarily stored in the middle chromosome $r2$ (*solution_buffer*). The length of $r2$ is $n - i$.

4. The $r2$ from Step 3 is combined to the rear of *child1*.

5. The two-point crossover operations are repeated for *parent2* as in the Steps mentioned above.

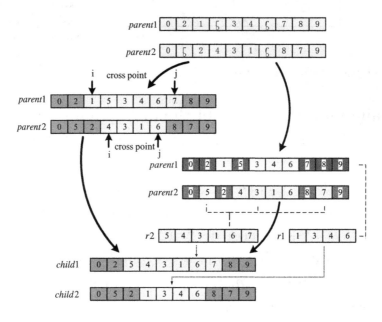

Figure 6.4   Two-point crossover.

The crossover algorithm is shown in Algorithm 5. In Algorithm 5, Step 1 initializes the new population which contains two individuals. Steps 2 to 9 form the loop guaranteeing that the produced new individuals are distinct. Step 5 invokes Algorithm 6, which presents a normal crossover procedure. Algorithm 7 ensures that the crossover operation gets a feasible individual without violating the precedence constraint among the DAG graph. In algorithm 7, Step 1 randomly selects two crossover points. Steps 2 to 4 are initial phases. Steps 6 to 9 compose the loop which duplicates the range before crossover point. Steps 10 to 14 compose the loop which shift the genes left after Steps 6 to 9 while satisfying the constraints among the parallel tasks. The time complexity of crossover algorithm is $O(n \cdot \log N)$, where N is the population size and n is the workflow size.

Input: Two parent individuals *parent*1 and *parent*2.
Output: Two child individuals *child*1 and *child*2.
1 $children \leftarrow \emptyset$;
2 **while** $i \in (1...POPULATION\_SIZE)$ **do**
3   **if** $random([0,1]) \leq CROSSOVER\_RATE$ **then**
4   Select two solutions randomly from the solution space as a tuple (*parent*1, *parent*2);
5   Invoke Algorithm Crossover1 with parameters (*parent*1, *parent*2) to produce two random solutions (*child*1, *child*2);
6   Add feasible solution *child*1 to the set *children*;
7   Add feasible solution *child*2 to the set *children*;
8   **end**
9 **end**

Algorithm 5: The Crossover algorithm

Input: two parent individuals *parent*1 and *parent*2.
Output: two child individuals *child*1 and *child*2.
1 $child1 \leftarrow$ Crossover2(*parent*1, *parent*2);
2 $child2 \leftarrow$ Crossover2(*parent*2, *parent*1);

Algorithm 6: The crossover1 algorithm

6.4.6 Mutation

In general, mutation operation includes simple, uniform, boundary, Gaussian, and non-uniform mutation. To escape from local optima, mutation operations are used in order to explore better solutions to some extent. As mentioned above, *URank* can maintain the topological order of the workflow tasks, i.e., the

Input: two parent individuals *parent*1 and *parent*2.

Output: a child individual *child*.

1 Obtain tuple (i, j), each element of which is randomly selected in the range of $[1 \leq i < j \leq N]$;

2 **for** *all* $k \in (1, ..., i-1, j+1, ..., N)$ **do**

3   task($child_k$) $\leftarrow$ task($parent1_k$));

4 **end**

5 $m \leftarrow 0$;

6 **for** *all* $[k \in (1, ..., N)] \wedge [task(parent2_k) \notin tasks(child)]$ **do**

7   task($solution\_buffer_m$) $\leftarrow$ task($parent2_k$);

8     $m \leftarrow m+1$;

9 **end**

10 **for** *all* $k \in (i, ..., j)$ **do**

11   **while** *all solution* $\in$ *population* **do**

12   task($child_k$) $\leftarrow$ task($solution\_buffer_{k-i+1}$);

13   **end**

14 **end**

Algorithm 7: The Crossover2 algorithm

precedence constraints among tasks. As illustrated in Figure 6.5, the mutation strategy used in this chapter is described as follows: two positive integers i and j are randomly selected while satisfying the condition $0 \leq i < j \leq n - 1$. When the *URank* of the selected tasks t_i and t_j (corresponding to genes i and j in the candidate chromosome, respectively) meet the inequality $URank(t_i) < URank(t_j)$, the genes i and j are swapped and a new chromosome is generated.

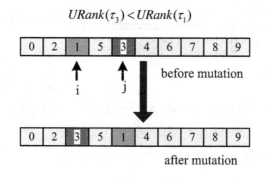

Figure 6.5 Mutation.

As shown in Algorithm 8, the mutation helps the candidate solution to escape from a local optimum. In Algorithm 8, Steps 1 to 5 make each new candidate solution mutate once. Steps 2 to 4 control the degree of mutation which depends on the MUTATION_RATE. Step 6 randomly selects two mutation points. Step 8 guarantees that the entire mutation conforms to the precedence constraints among parallel tasks. The time complexity of mutation algorithm is $O(N)$, where N is the population size.

Input: all children individuals.
Output: new children individuals.
1 **for** *all solution $\in$ children* **do**
2   **if** *random($[0,1]$) $\leq$ MUTATION_RATE* **then**
3   Go to Step 7 and perform the mutate operation with parameter *solution*;
4   **end**
5 **end**
6 Obtain tuple (i, j), each element of which is randomly selected in the range of $[1 \leq i < j \leq N]$;
7 Swap task i and task j in the chromosome;
8 Check whether the *URank* of t_i and t_j satisfies the priority constraint relation of task set.

Algorithm 8: The Mutation algorithm

6.4.7 The Main Algorithm

The BOGA pseudocode is shown in Algorithm 9. In the population initialization, the priority queue is established with several famous strategies such as HEFT [146], HCPFD [59], PEFT [8], and that is used to create a feasible precedence queue. This operation also improves the convergence speed of the algorithm.

Input: A DAG $G = <V, E>$ and a set PE of DVS available processors.

Output: A pareto front which contains a set of schedule for G on PE.

1 Initialize A *population* of size POPULATION_SIZE;

2 Evaluate the quality of *population* by the fitness function (Eqs. (6.8) and (6.9));

3 GENERATION ← 1

4 **while** $GENERATION \leq GENERATION\_MAXIMUM$ **do**

5   Select two solutions randomly from the *population* as *parent1* and *parent2*;

6   Crossover *parent1* and *parent2* to obtain *children1* and *children2*;

7   Mutate *children1* and *children1* separately;

8   Evaluate the quality of *children1* and *children2* by the fitness function (Eqs. (6.8) and (6.9));

9   Append qualified individuals of *chilren1* and *chilren1* to *population*;

10   Update individuals in the *population*;

11   GENERATION ← GENERATION+1;

12 **end**

Algorithm 9: The BOGA algorithm

In BOGA, Step 2 calculates the entire energy consumption and POF for each individual when it is evaluated. Steps 5—11 are repeated until the iteration circle reaches the specified maximum. Two parent individuals are randomly selected and the selected individuals are guaranteed to be different in Step 5. After Step 6, two new individuals are generated. In order to prevent the early convergence and escape from the local optima, mutation operation is devised for parallel tasks in Step 7 which is employed to expand the solution space of this chapter. For the new generated individuals, only the one which dominates its parent is selected for the next population. When BOGA proceeds to Step 10, the size of the current population is twice $(2N)$ that of the initial population. Inspired by the fast and elitist multi-objective genetic algorithm reported in [7], the $2N$ individuals are processed with fast non-dominated sorting and crowding distance calculation. Then, each individual is assigned with *rank* values and crowding distance, respectively. During the process of selection, the individual with a low *rank* has a higher priority to be an elite. Next is the one with a large crowding distance value in the same *rank* level. This process repeats until the size of the elite individuals reaches N. The whole algorithm runs until the iterations reach the upper bound. The latest individuals N elites are the Pareto set.

For a workflow with n nodes, a is the number of directed edges allowing the p processors to schedule the n tasks. The complexity of the BOGA algorithm is $O(n \log n + (a + n)p)$.

6.5 PERFORMANCES EVALUATION

In this section, experiments are conducted to verify the effectiveness of BOGA. In the following descriptions, two performance metrics are firstly introduced. Then, the configuration of the experiment parameters is shown. Real-world and randomly generated application graphs are used to assess the performance of the proposed algorithms. Lastly, the experiment results are analyzed.

6.5.1 Performance Metrics

The fitness functions, including ECR and POF expressed in Eqs. (6.8) and (6.9) respectively, which clearly show the pareto front in figures, are exploited to evaluate the performance of the presented BOGA algorithm and comparison algorithms.

6.5.2 Experimental Setting

In this section, several benchmarks used in workflow model are exploited to test BOGA. These workflows were designed to resolve several particular parallel numeric computation problems. These benchmarks include parallel Gauss-Jordan algorithm to

TABLE 6.3 Selected workflow models

Workflow	# Nodes	Reference	Note
T1	15	[149]	Gauss-Jordan Algorithm
T2	14	[149]	LU decomposition
T3	16	[158]	Laplace Transform

solve systems of equations [148], parallel LU decompositions [148], and discrete Laplace transformation [158]. Detailed parameters are shown in Table 6.3.

Before making a comparison with multi-objective differential evolution (MODE) and MOHEFT algorithms, a brief introduction is firstly given. The parent and several other individuals randomly selected are combined to generate new candidate solutions. The parent is replaced by a candidate when it has a worse fitness value. It is a great selection strategy that often surpasses other classic evolution algorithms. In addition, the concept of dominance is used to preserve a uniformly spread front of non-dominated solutions. Then, the population size grows. The MODE truncates to prepare it for the next round of iterative evolution operations. While the MOHEFT algorithm is based on HEFT, a solution is created by iteratively mapping tasks onto appropriate resources. Its main objective is to minimize the entire completion time of each task. When multiple objectives are simultaneously considered, the goal changes to calculate the set of tradeoff solutions. In MOHEFT, several solutions are allowed to generate simultaneously instead of creating a single solution. In addition, the tasks are allowed to map onto other resource which has a late finish time when the tradeoff of this resource is between the considered objectives.

The experiments of the present chapter are carried out using a workstation at the National Supercomputing Center in Changsha. It is equipped with an Intel Core i3-540 dual-core CPU, 8 GB DRAM, and 500 GB hard disk. The machine runs with Windows 7 (64-bit OS) SP1. The presented algorithm is implemented in an open-source framework jMetal [48], which is an object-oriented Java-based framework aimed at the development, experimentation, and study of metaheuristics for solving multi-objective optimization problems. The earliest version of jMetal is used to implement the multi-objective optimization problem of continuous functions. In the past 2 years, several researchers developed it to solve discrete optimization problem. The workflow scheduling problem developed in this chapter refers to a discrete and combinatorial multi-objective optimization problem in which each task has precedence constraint. In order to further exhibit BOGA performance, it was compared to the performance of MODE [133]. MODE introduces a different evolution idea to solve workflow scheduling in the global grid. Each algorithm iterates 100 times. The benchmark used in this chapter involves three different types of workflows. Under these workflows in the benchmark, the minimum extent in each objective is used to evaluate the quality of the proposed algorithms.

6.5.3 Real World Application Graphs

Task graphs from four kinds of real-world applications are used in these experiments: Gauss-Jordan [149], LU decomposition [149], Laplace transformation [158], and molecular dynamic code [72].

6.5.3.1 Three Kinds of Classic DAG Graphs

According to the experimental parameters in Table 6.3, three heterogeneous processors configurations are used to schedule the three different types of workflow, i.e., T1, T2, and T3, respectively. The performance metric used for comparison are ECR and POF. The communication to computation ratio (CCR) of a DAG is used to depict the feature of the graph. It is either a communication-intensive or a computation-intensive application. When the value of CCR is higher, DAG is a communication-intensive task graph. On the contrary, the lower the CCR value is, the more computation-intensive the DAG is. The comparisons of the three workflows under algorithms BOGA, MODE, and MO-HEFT are illustrated as Figures 6.6, 6.7, and 6.8, respectively, when CCR is set to 1.

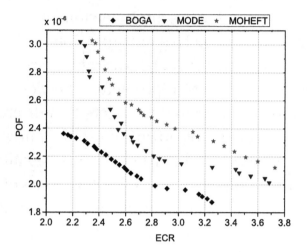

Figure 6.6 Comparisons of Gauss-Jordan (the number of processors equals 3 with CCR = 1.0).

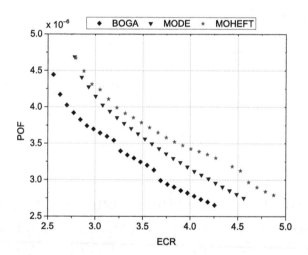

Figure 6.7 Comparisons of Laplace (the number of processors equals 3 with CCR = 1.0).

Figure 6.6 shows the Gauss-Jordan workflow under the three kinds of algorithms. MODE, based on the differential evolution idea, finds a better Pareto front compared with MOHEFT. As the value of ECR is above 2.62, the POF of MOHEFT significantly increases. For BOGA, a fine initial population is obtained when

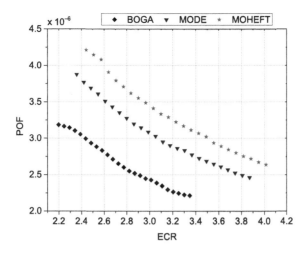

Figure 6.8 Comparisons of LU (the number of processors equals 3 with CCR = 1.0).

the famous list schedule algorithms such as HEFT, HCPDF, and PEFT are employed. During each BOGA iteration, the solutions in the population proceeds with non-dominated sorting and crowding distance operations to get nondominated fronts. Then, specific one-point crossover and mutation operators for parallel tasks are employed so that the generated individuals are invalid and they reserve the most elitist individuals from its parent generations. Therefore, BOGA obtains the best Pareto front in comparison with MODE and MOHEFT. In addition, the classical algorithms used in the beginning of BOGA greatly improve the efficiency of initial population constructing, which has no impact on the time complexity of BOGA.

Likewise, as shown in Figures 6.7 and 6.8, BOGA consistently performs well. I_H^- and $I_{\varepsilon+}$ of BOGA are the minimum of the three algorithms. The difference is that T1, T2, and T3 are three different kinds of workflow in structure. The average in-degree (or the average out-degree) of T1, T2, and T3 are not the same because the task set have different parallel degrees. When the number of workflow is almost the same, the ECR and POF of the obtained Pareto front under the three existing algorithms have several differences.

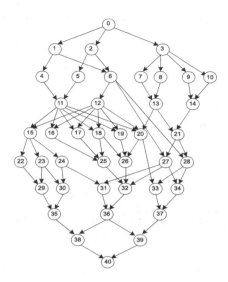

Figure 6.9 A molecular graph.

6.5.3.2 Molecular Dynamic Code

DAG, extracted from the molecular dynamic code presented in [72], is used to evaluate the performance of the scheduling algorithms in this experiment. DAG is shown in Figure 6.9.

The maximum out-degree in this specific DAG is 7 as shown in Figure 6.9. The parameter of the maximum out-degree in a DAG has a great impact to the performance of algorithms. The parallelism degree largely depends on these parameters to some extent. When the value of the maximum out-degree is large enough, further increasing the number of processors has less effect in improving performance.

Figure 6.10 shows the performance of the three algorithms when the molecular dynamic code DAG graph is scheduled on six different processor with CCR = 1.0. These results suggest that MODE and MOHEFT effectively performs to explore feasible solutions. The BOGA presented in this chapter delivers more consistent performance in finding better Pareto front solutions. In Figure 6.10, the ECR increases as the POF decreases. This result once again verifies that BOGA is able to strike a fine balance between energy consumption and system reliability.

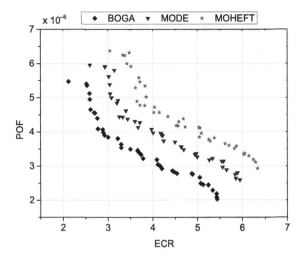

Figure 6.10 Comparisons of molecular graph (the number of processors equals 6 with CCR = 1.0).

6.5.4 Randomly Generated Application Graphs

Without loss of generality, randomly generated task graphs are used to assess the performance in these experiments. Yuming Xu *et al.* [165] developed a random graph generator, which specifies different graphs of DAG input parameters (i.e., input degree, out-degree, the number of DAG, the level of DAG, the value of CCR, and so on) by the users. With this task generator, a large set of random task graphs are generated with different characteristics. All of these graphs are scheduled on heterogeneous computing systems. To evaluate the performance of BOGA, MODE, and MOHEFT under different parameters involving various numbers of computing processors, DAG task different numbers and different CCR values are compared.

In this chapter, the input parameters used in the experiments is as follows: the number of task is 100, the levels of tasks ranges from 7 to 57, the maximum DAG in- and out-degrees are 47 and 57, respectively, and the average in-degree is 7.81. With respect to the number of CCRs, three kinds of values are specified–0.5, 1.0, and 5.0, respectively.

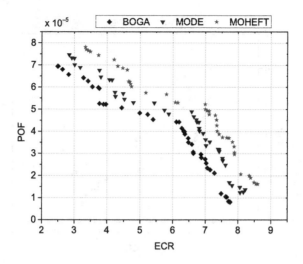

Figure 6.11 Comparisons of randomly generated DAG graph (the number of task graphs equals to 100, the number of processors equals 6 with CCR = 0.5).

Figures 6.11 to 6.13 show the comparisons of the three algorithms which run on three heterogeneous processors under different CCRs. As shown in Figure 6.11, the application is computation intensive with CCR = 0.5 wherein the solutions explored by BOGA are significantly better than MODE and MOHEFT. When the user needs a lower POF with a modest energy consumption, the obtained Pareto front produced by BOGA

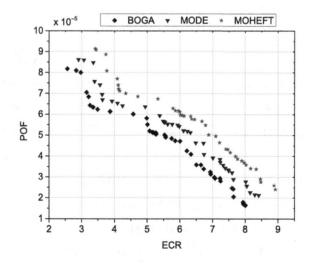

Figure 6.12 Comparisons of randomly generated DAG graph (the number of task graphs equals to 100, the number of processors equals 8 with CCR = 1.0).

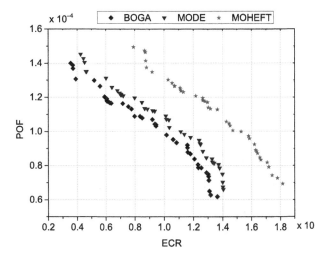

Figure 6.13 Comparisons of randomly generated DAG graph (the number of task graphs equals to 100, the number of processors equals 6 with CCR = 5.0).

offers a large number of candidate solutions. For Figure 6.12, the application is modest in computation and communication. The POFs and ECRs explored by BOGA, MODE, and MOHEFT maintain the same trend as CCR = 0.5. BOGA outperforms the other two algorithms in finding a better set of feasible solutions. As illustrated in Figure 6.13, when CCR is equal to 5, DAG is a communication-intensive graph and BOGA performs consistently better over MODE and MOHEFT. The fine performance of BOGA is attributed to the fact that the mechanisms of the initial population and selection in BOGA are more intelligent.

6.6 SUMMARY

The data center provides users with different quality of service such as maximum completion time, energy-saving effect, and system reliability to satisfy their different requirements. This chapter presents the dual objectives of energy saving and system reliability for the workflow with precedence constraints in heterogeneous systems. BOGA is developed to obtain a fine Pareto front. In the beginning of BOGA, several well-known, efficient strategies are adopted to the initial population without introducing additional computation complexity. For the selection operation, the two objectives of low energy consumption and low POF are used to determine the non-dominated solutions. The specific one-point crossover is designed to generate new individuals without violating the precedence constraint. The selected solutions that fulfill the *URank* constraint perform a particular mutation operation, thereby helping the BOGA to escape from the local optima and explore a new solution space. While entering the next iteration, two operations, namely, non-dominated sorting and crowding distance, are performed, thereby reserving the diversity of individuals. Unlike in modified MODE and MOHEFT, three kinds of workflows, i.e., Gauss-Jordan, parallel LU decomposition and discrete Laplace transform, and real-world application graphs (i.e., molecular dynamics code graph and randomly generated application graphs), are exploited to assess performance.

CHAPTER 7

Interconnection Network Energy-Aware Scheduling Algorithm

With the development of social economy and computing infrastructure, heterogeneous systems based on multi-core (CPU) and many-core (GPU) processors have attracted more and more attention. Many large-scale computing intensive scientific workflow applications have been deployed on such systems. However, it is still a challenging problem to reduce systems energy consumption and improve performance under parallel application deadline constraints. This chapter first presents the computing node network energy consumption problem of the most famous fat-tree interconnection networks for low communication to computation ratio parallel application. Then, we present a heuristic network energy-efficient workflow task scheduling algorithm, which includes task level computing, task subdeadline initialization, task dynamic adjustment, and a data communication optimization scheme among tasks.

7.1 INTRODUCTION

With the advent of multi-core (CPU) and many-core (GPU) processors, large-scale heterogeneous systems (such as

DOI: 10.1201/b23006-7

Supercomputers Sierra, Perlmutter, Summit, Tianhe-2A, Selene, Marconi-100, Titan, Piz Daint, and so on) have risen as a primary and high-effective computing infrastructure for high-performance applictions [1, 24, 113, 141, 163]. For example, work [113] successfully completed a nonlinear AWP-ODC scientific applications on 4,200 Kepler K20X GPUs on Oak Ridge National Laboratory's Supercomputer Titan, which simulates a 7.7M earthquake on the southern San Andreas fault. Paper [141, 163] implemented large-scale high-order CFD (Computational Fluid Dynamics) simulations on the Tianhe-1A supercomputer, which adopted hybrid MPI+OpenMP+CUDA programming model to realize the parallelism of 1024 computing nodes.

These heterogeneous systems composed of millions of cores will inevitably consume a lot of electrical power and incur high operation costs. This is a key challenge for the development and management of large-scale heterogeneous computing systems that provide high energy-efficiency solutions.

Several strategies have been presented to reduce energy consumption on large-scale heterogeneous systems:

- Hardware level. These are many techniques such as dynamic voltage and frequency scaling (DVFS), resource hibernation.

- Architecture level. It can design and optimize a simple and energy-efficient heterogeneous framework.

- System resource management level. It comprehensively considers system computing resources and user parallel applications to optimize the resource scheduling [46, 86, 153].

However, most of the classical resource management approaches of heterogeneous systems focus on the computing processors, memory, storage, and ignore the energy consumption of high-speed interconnection networks. In fact, as reported in [150], the energy consumption of modern high-speed interconnection networks has increases to 10–15 percent of the total energy of the systems [141].

On the other hand, high-performance scientific parallel applications are usually computation intensive and compose of thousands of precedence constrained tasks. Examples include earthquake engineers Broadband, seismic hazard calculations via Cybershake, weather prediction, and bioinformatics Epigenomics [3,29,32,147]. Most high-performance workflow applications have low communication to computation ratio (CCR), which means low data communication between tasks, and resulting in low utilization of heterogeneous computing systems high-speed interconnection network [25,141]. Therefore, the systems interconnection network resources may remain idle at some time intervals. In addition, due to the limitation of workflow application tasks precedence, most parallel application tasks cannot be executed when the computing nodes are available [141]. This phenomenon also improves the idle rate of computing nodes and network interconnection resources. We can dynamically switch on/off these idle systems resources to reduce energy consumption. Thirdly, some scientific parallel applications are expected to be completed its execution before a certain time (or deadline), such as weather prediction [4,9,111,141,147]. Motivated by the above challenges, this chapter mainly focuses on the problem of efficiently scheduling deadline-constrained workflow high-performance application on heterogeneous systems to achieve optimal computing and network interconnection resources energy consumption.

7.2   HETEROGENEOUS SYSTEMS

This section introduces the architecture of heterogeneous systems computing node and high-speed interconnection network, scientific parallel application workflow DAG model, and resources energy consumption model used in this chapter [141].

7.2.1   Computing Nodes and Fat-Tree Networks

Most heterogeneous computing systems are mainly consisted of CPU+GPU computing nodes and high-speed interconnection networks. For example, the computing nodes of Summit super-computer are equipped with two IBM Power9 CPUs, and six NVIDIA Tesla V100 GPUs [1]. These nodes are linked by a dual-rail Mellanox EDR InfiniBand high-speed interconnection network [100].

Figure 7.1 is a schematic diagram of the heterogeneous systems fat-tree interconnection network topology [141]. Some computing nodes have 3 multi-core CPUs, 1 many-core GPUs; Some have 2 multi-core CPUs, 1 many-core GPUs, and some have only 2 multi-core CPUs. These computing nodes are connected by a network routing chip (NWR) [162]. All of the computing node CPUs/GPUs have been deployed with intelligent energy consumption management techniques that can automatically control them in execution and sleeping states [153]. The NWRs are responsible for packet switching and routing among computing nodes, and the aggregated communication among computing node CPUs and GPUs is carried out through node local backplane [141]. Here, it is assumed that the network routing chip can dynamically power on/off with negligible overhead.

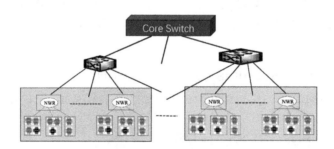

Figure 7.1   A heterogeneous computing systems fat-tree architecture.

This chapter assumes that systems consist of a set of computing nodes $CN = \{CN_1, CN_2, \cdots, CN_n\}$. Each computing node is denoted as $CN_j = \langle cg_j, m_j, nwr_j \rangle$, where we has m_j multi-core CPUs and many-core GPUs. The network routing chip associated with node CN_j is name as nwr_j. This chapter does not consider the aggregation of heterogeneous systems and core layer interconnection network resources, because they are always shared by the systems [141].

7.2.2 Scientific Application Workflow

In general, high-performance scientific parallel applications usually consist of a set of precedence constrained tasks. Some of them are suitable for process by multi-core CPUs and others are suitable for many-core GPUs [141]. These tasks have control and data communication among them. The extended DAG model of such application workflow is derived from Section 1.3.2 and expressed as $G = \langle T, A, W, D, dl \rangle$ [161]. Here, T is a set of $|T|$ precedence constrained tasks, whereby t_i^c can be scheduled to any available multi-core CPU heterogeneous systems and t_i^g is suitable for many-core GPU; $w_i \in W$ denotes task t_i's computation time on corresponding computing resource [141]. $a_{i,j} \in A$ represents tasks control and data communication constraint, and $d_{i,j} \in D$ is the data communication time between task t_i and t_j. In this system environment, the data communication time $d_{i,j}$ can be neglected as tasks t_i and t_j are scheduled on the same computing node. dl is the parallel application workflow deadline [141]. Figure 7.2 shows a simple extended DAG model.

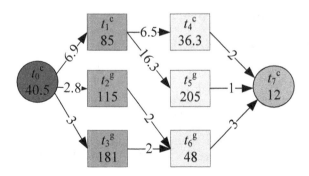

Figure 7.2 The illustration of extended DAG model.

7.2.3 Energy Consumption Model

This chapter considers the energy consumption of computing node that consist of multi-core CPU, many-core GPU, and network routing chip NWR. The basic concepts of power can be seen in Section 5.2.4. For a day, task t_i is assigned to the kth CPU/GPU of the jth computing node CN_j, the energy consumption can be described as

$$
\begin{aligned}
E[cg_j^k] &= \Phi(cg_j^k) \times w_i + \Phi_s(cg_j^k) \times (3600 - w_i), \\
&= \Phi_d(cg_j^k) \times w_i + \Phi_s(cg_j^k) \times 3600.
\end{aligned}
\tag{7.1}
$$

Here, we also assume that the power of computing node network routing chip NWR is $\Phi(nwr_j)$, and the actual NWR work time in a day is $y(nwr_j)$. Therefore, the energy consumption of the computing node NWR is

$$
E[nwr_j] = \Phi(nwr_j) \times y(nwr_j).
\tag{7.2}
$$

In this way, the energy consumption of system computing node CN_j is the sum of CPUs/GPUs and NWR, which can be denoted as,

$$
E[CN_j] = \sum_{k=1}^{m_j} E[cg_j^k] + E[nwr_j].
\tag{7.3}
$$

Therefore, the energy consumption of the heterogeneous system is the total of all computing nodes and other system resources energy consumption, such as the core layer network and cooling system [141]. This chapter does not consider optimizing the energy consumption of other resources. It can be defined as follows,

$$
E[system] = \sum_{j=1}^{n} E[CN_j] + E[other].
\tag{7.4}
$$

7.3 INTERCONNECTION ENERGY AWARE SCHEDULING PROBLEM

This section considers the problem of scheduling high-performance parallel application tasks (Figure 7.2) on two computing nodes as shown in Figure 7.3. Here, the $cg^1, \cdots, cg^4$ are the multi-core

CPUs and cg^5 and cg^6 are many-core GPUs. In the current system execution state, five tasks t_0^c, t_1^c, t_2^g, t_3^g, and t_4^c have been scheduled on cg_1^1, cg_1^2, cg_1^5, cg_1^6, and cg_1^3 of computing node CN_1, respectively. Now, we consider schedule task t_5^g, one solution is assigning t_5^g to GPU cg_1^5 that its execution start time is 155.5. In this case, the data communication between tasks t_0^c, t_1^c, and t_5^g will become zero because they are on the same computing node CN_1. Another option is assigning t_5^g on the GPU cg_2^5 of computing node CN_2; Here, its execution start time will become 141.8, which is earlier than the first scheme [141]. However, in this circumstance, tasks t_1 and t_5 have data communication $d_{1,5}=16.3$ through NWR nwr_1 and nwr_2, which will result in additional communication overhead and computing node network routing chip NWR energy consumption.

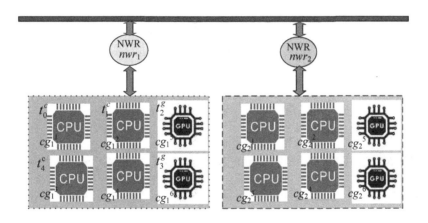

Figure 7.3 Task scheduling across computing nodes problem.

The second problem is how to arrange task communication to further save network routing chip NWR energy consumption. Figure 7.4(a) depicts a computing node NWR data communication occupancy schedule [141]. There is a communication request with the data communication time $d_{i,j}=50$, and its earliest and latest start time are 70 and 100, respectively. Here, we can give three schemes to arrange this data communication request, which are described in Figure 7.4(b), Figure 7.4(c), and Figure 7.4(d).

Accordingly, the data communication start time may be its earliest start time, occupied time period and latest start time. Intuitively, Figure 7.4(c) shows a better data communication scheme. This is because its data communication time is shared with others, and the solution can reduce network routing chip NWR energy consumption by turning it off [141]. Therefore, we mainly focus on task scheduling that can simultaneously optimize both system energy consumption and parallel application performance.

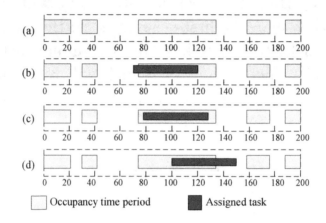

Figure 7.4 Network routing chip data communication time.

7.4 NETWORK ENERGY-EFFICIENT WORKFLOW SCHEDULING STRATEGY

This section presents a heuristic list-based network energy-efficient workflow task scheduling algorithm, which we name it as NEEWS. This algorithm mainly consist of four phases: task level computing, task subdeadline initialization, dynamic adjustment, and a data communication optimization algorithm [141].

7.4.1 Task Level Computing

The NEEWS algorithm attempts to schedule high priority dependent tasks on the same computing node of the systems, which can save the data communication between tasks as much as possible. In addition, this chapter mainly focuses on scientific computation intensive parallel application with little communication between tasks [141]. Therefore, we modify the classical workflow DAG model task top level t_level [43], which does not consider workflow edges data communication. In general, the t_level can be computed from the *entry* task to task in the workflow topological order, and is expressed as

$$\begin{cases} t\_level(t_{entry}) = w_{entry}, \\ t\_level(t_i) = w_i + Max_{k \in prev(t_i)}\{t\_level(t_k)\}. \end{cases} \quad (7.5)$$

The task rank value will be used in the following workflow tasks scheduling.

7.4.2 Subdeadline Initialization

The workflow is constrained by its deadline dl, and we present a deadline distribution technique that is divided into two phases: subdeadline initialization and dynamic adjustment [141]. In the first phase, we assign a initialization subdeadline to each individual task according to the task computation time. We define the subdeadline division value as $dlel$, which is defined as,

$$\begin{cases} dlel(t_{exit}) = t\_level(t_{exit}), \\ dlel(t_i) = Min_{k \in next(t_i)}\{dlel(t_k) - w_k\}, \end{cases} \quad (7.6)$$

where the $dlel$ can be computed from the *exit* task to task in the workflow DAG reversed topological order. Moreover, task subdeadline dl_i can be computed by

$$dl_i = \frac{dlel(t_i)}{t\_level(t_{exit})} \times dl. \quad (7.7)$$

The pseudo-code of parallel application deadline division initialization is shown in Algorithm 10. Table 7.1 lists the tasks t_level and $dlel$, and the corresponding initialization subdeadline dl_i for Figure 7.2. Here, the deadline is set as $dl = 400$.

Input: Workflow DAG

Output: Each tasks subdeadline dl_i

1   Construct a list of tasks in topological order;

2   $t\_level(t_{entry}) \leftarrow w_{entry}$;

3   **for** *each task t_i in list* **do**

4   Compute task $t\_level(t_i)$ using Eq. (7.6);

5   **end**

6   Construct a list of tasks in reversed topological order;

7   $dlel(t_{exit}) \leftarrow t\_level(t_{exit})$;

8   **for** *each tasks t_i* **do**

9   Compute task $dlel(t_i)$ using Eq. (7.7);

10   Compute task subdeadline dl_i using Eq. (7.8);

11   **end**

Algorithm 10: Deadline Initialization.

7.4.3   Dynamic Adjustment

In the subdeadline initialization phase, we do not consider the tasks data communication. In fact, data communication inevitably exists in scientific parallel applications. For instance, in Figures 7.2 and 7.3, there is the data communication $d_{1,5}$ through computing node NWRs nwr_1 and nwr_2 when task t_5^g is scheduled on the computing node CN_2 GPU cg_2^5. In such case, the $t\_level(t_5)$ is increased to 346.8, and the unscheduled tasks t_level, $dlel$, and dl_i may also change as it. In view of this, we present a dynamic adjustment approach based on the task subdeadline initialization, which is outlined in Algorithm 11 [141]. Table 7.2 lists the adjusted unscheduled tasks t_level and $dlel$, and the subdeadline dl_i as the task t_5^g is scheduled.

TABLE 7.1   Tasks t_level, $dlel$, and dl_i of Figure 7.2

$Task$	t_0	t_1	t_2	t_3	t_4	t_5	t_6	t_7
t_level	40.5	125.5	155.5	221.5	161.8	330.5	269.5	342.5
$dlel$	40.5	125.5	282.5	282.5	330.5	330.5	330.5	342.5
dl_i	47.3	146.6	330	330	386	386	386	400

Input: Application workflow partial information
Output: Adjusted unscheduled tasks value
1 Get the unscheduled tasks topological list;
2 **for** *each unscheduled tasks* t_i **do**
3   Compute task $t\_level(t_i)$ using Eq. (7.6);
4 **end**
5 Get the unscheduled tasks reversed topological list;
6 $dlel(t_{exit}) \leftarrow t\_level(t_{exit})$;
7 **for** *each tasks* t_i **do**
8   Compute task $dlel(t_i)$ using Eq. (7.7);
9   Compute task subdeadline dl_i using Eq. (7.8);
10 **end**

Algorithm 11: Adjustment Subdeadline

TABLE 7.2 The adjusted unscheduled tasks for Figure 7.2

$Task$	t_0	t_1	t_2	t_3	t_4	t_5	t_6	t_7
t_level	40.5	125.5	155.5	221.5	161.8	346.8	269.5	358.8
$dlel$	40.5	125.5	282.5	282.5	330.5	346.8	346.8	358.8
dl_i	47.3	146.6	330	330	386	386.6	386.6	400

7.4.4 Data Communication Optimization Algorithm

This section defines some important task scheduling attributes. We let *EES* denotes the earliest execution start time, and *EEF* is the earliest execution finish time [141]. Here, we also assume that task t_i assigning to the kth multi-core CPU of computing node CN_j is denoted as cg_j^k. Thus, the task's earliest execution finish time can be expressed as,

$$EEF(t_i, cg_j^k) = EES(t_i, cg_j^k) + w_i. \qquad (7.8)$$

Here, the task t_i's earliest execution starting time is restricted by computing resource availability and task precedence constraints. We describe it as follows,

$$\begin{cases} EES(t_i, cg_j^k) \geq Avail(cg_j^k), \\ EES(t_i, cg_j^k) \geq Max_{x \in prev(t_i)}\{EFT(a_{x,i})\}, \end{cases} \qquad (7.9)$$

where $EFT(a_{x,i})$ is the edge $a_{x,i}$ communication finish time. We describe it as the sum of the edge $a_{x,i}$ start communication time $EST(a_{x,i})$ and the data communication time $d_{x,i}$. That is,

$$EFT(a_{x,i}) = EST(a_{x,i}) + d_{x,i}. \tag{7.10}$$

When the parent task t_x scheduled computing resource r and cg_j^k are in the same systems computing node, the data communication time $d_{x,i}$ between them will become zero [141]. That is to say: $EST(a_{x,i})=EEF(t_x, r)$. Otherwise, the data communication $a_{x,i}$ must be carried out between nodes, such as computing nodes CN_j and CN_y, and will lead to network communication and energy consumption. In addition, the main goal of the study of this chapter is to save network routing chip NWR energy consumption [141]. The typical example shown in Figure 7.4 demonstrates that a better data communication between tasks scheme can significantly reduce network routing chip energy consumption [141].

For each system computing node CN_y, we define a network routing chip NWR nwr_y communication occupancy time linked list ll_y. Each item of ll_y records the communication start and end time points of the NWR discrete period, such as Figure 7.4(a). Intuitively, the earliest data communication start time of the DAG edge $a_{x,i}$ is $EST(a_{x,i})=EEF(t_x, r)$. Then, the execution finish time of task t_i is constrained by its subdeadline dl_i. Thus, the latest DAG edge $a_{x,i}$ data communication start time $LST(a_{x,i})$ can be expressed by

$$LST(a_{x,i}) = dl_i - w_i - d_{x,i}. \tag{7.11}$$

This chapter tries to find an optimal data communication start time $EST(a_{x,i})$ between nodes CN_y and CN_j with minimum network routing chip NWR energy consumption. For a given edge data communication start time $EST(a_{x,i})$ and communication time $d_{x,i}$, we can query the NWR nwr_j occupancy time linked list ll_j and obtain its non-shared part of the communication time $NSCT(a_{x,i})$, such as in Figures 7.4(b) and 7.4(d). When the data communication is shared with others, such as in Figure 7.4(c), the $NSCT(a_{x,i})$ will become 0. Therefore, the application DAG edge $a_{x,i}$ energy consumption of node NWR nwr_j is

$$E(a_{x,i}, nwr_j) = \Phi(nwr_j) \times NSCT(a_{x,i}). \qquad (7.12)$$

This energy consumption computation approach is derived from Eq. (7.4) and tries to achieve optimal solution. The edge optimization communication start time $EST(a_{x,i})$ search strategy is described in Algorithm 12 [141].

Input: Edge $EST(a_{x,i})$, $LST(a_{x,i})$, $d_{x,i}$, and computing nodes ll_y, ll_j
Output: Edge $EST(a_{x,i})$, $NWREnergy$
1 $NWREnergy \leftarrow MaxEnergy$;
2 **for** $i \leftarrow EST(a_{x,i}); i < LST(e_{x,i}); i++$ **do**
3   Give edge data communication start time as i;
4   Compute $E(a_{x,i}, nwr_y)$ using Eq. (7.13);
5   Compute $E(a_{x,i}, nwr_j)$ using Eq. (7.13);
6 **end**
7 **if** $NWREnergy > E(a_{x,i}, nwr_y) + E(a_{x,i}, nwr_j)$ **then**
8      $EST(a_{x,i}) \leftarrow i$;
9      $NWREnergy \leftarrow E(a_{x,i}, nwr_y) + E(a_{x,i}, nwr_j)$;
10 **end**

Algorithm 12: The optimization edge communication start time search algorithm

7.4.5 The Heuristic Network Energy-Efficient Workflow Scheduling Algorithm

The pseudo-code of the heuristic network energy-efficient workflow scheduling algorithm (NEEWS) is shown in Algorithm 13 [141]. First, we use Algorithm 10 to compute workflow task scheduling list attributes: top level t_level and subdeadline dl_i. Then, the tasks with minimum t_level are selected to search the most suitable solution one by one [141]. The variable $TaskE$ is initialized as the maximum energy consumption in Step 4. From Step 5 to Step 26, we try to find an optimal computing resource cg_j^k with minimum task energy consumption by searching all computing nodes CN_j. Here, Steps 6–15 use Algorithm 12 to compute task t_i and its parent task data communication scheme with optimization energy consumption. The condition $TypeCond$ means that t_i^c is only scheduled on a multi-core CPU and t_i^g is on a many-core GPU. After that, an optimal computing resource is found through the whole systems computing nodes. At last, we assign task to the corresponding systems computing resource and update their information. We also use Algorithm 11 to adjust unscheduled task scheduling attributes.

The time complexity of workflow DAG scheduling algorithms is usually expressed in terms of task number $|T|$, total computing resource number $TRN=\sum_{j=1}^{n}\{CN_j \times m_j\}$. For each tasks, we test all computing nodes and its resources. Therefore, the time complexity of the network energy-efficient workflow scheduling algorithm is $O(|T| \times TRN)$ [141].

7.5 REAL-WORLLD APPLICATION PERFORMANCE EVALUATION

This section compares the heuristic network energy-efficient workflow task scheduling algorithm (NEEWS) with EATS [58], IC-PCP [4], and HEFT [146].

7.5.1 Experimental Setting

The experiments are conducted by a discrete event systems simulator like Tianhe-1, which consists of 40 computing nodes with 2 multi-core CPUs and a many-core GPU [141]. These computing nodes are connected by high-speed fat-tree networks. The performance comparison metrics chosen in this chapter are the system energy consumption (Eq. (7.5)). In order to achieve comprehensive insights into the effectiveness of the NEEWS algorithm, we compare these algorithms based on Montage and LIGO real-world workflow parallel applications [141].

On the other hand, the applications deadline constraints also have an important influence on the comparison metrics of these workflow scheduling algorithms. In this comparison experiments, the algorithm HEFT is a traditional and non-deadline-constrained strategy, and try to obtain the optimal schedule length for all algorithms [146]. Actually, the deadline-constrained workflow scheduling algorithms, such as NEEWS, EATS, and

Input: Workflow DAG and Computing Nodes
Output: The schedule solution

1 Compute each task attributes using Algorithm 10;
2 **while** *there are unscheduled tasks* **do**
3   Find a task t_i with minimum t_level;
4     $TaskE \leftarrow MaxEnergy$;
5   **for** *each computing nodes* CN_j **do**
6         $NetE \leftarrow 0$;
7   **for** *each tasks* $t_x \in prev(t_i)$ **do**
8   Task t_x is scheduled on CN_y;
9   Let $EFT(a_{x,i}) \leftarrow EEF(t_x, CN_y)$;
10   **if** $CN_j \neq CN_y$ **then**
11   Get edge $EST(a_{x,i})$, $NWREnergy$ using Algorithm 12;
12   Compute $EFT(a_{x,i})$ using Eq. (7.11);
13                 $NetE \leftarrow NetE + NWREnergy$;
14   **end**
15   **end**
16   **for** $k \leftarrow 1; k <= m_j; k++$ **do**
17   Compute $EEF(t_i, cg_j^k)$, $EES(t_i, cg_j^k)$ using Eq. (7.9), Eq. (7.10);
18   **if** $EEF(t_i, cg_j^k) < dl_i$ and $TypeCond$ **then**
19   Compute task energy consumption $E(t_i, cg_j^k)$ using Eq. (7.2);
20   **if** $TaskE > E(t_i, cg_j^k) + NetE$ **then**
21                     $TaskE \leftarrow E(t_i, cg_j^k) + NetE$;
22   Let cg_j^k as scheduled resource;
23   **end**
24   **end**
25   **end**
26   **end**
27   Assign task t_i to cg_j^k;
28   Update computing nodes information;
29     $t\_level(t_i) \leftarrow EEF(t_i, cg_j^k)$;
30   Adjust unscheduled tasks using Algorithm 11;
31 **end**

Algorithm 13: The pseudo-code of network energy-efficient workflow scheduling algorithm

IC-PCP, cannot successfully schedule all precedence constrained tasks of the application before a minimum deadline. Thus, in the following experiments, we let the applications' deadline based on the schedule length produced by HEFT [141].

7.5.2 Real-World Scientific Workflow

These experiments use two real-world performance evaluation workflow parallel application DAG: Montage (Astronomy, Figure 1.7) [69,112] and LIGO (Gravitational physics, Figure 1.10). The Montage is an astronomy parallel application used to construct mosaics of the sky (the details can be seen in Section 1.4) [69,112]. The LIGO is used in physics for detecting gravitational waves (the details can be seen in Section 1.4). Intuitively, the LIGO workflow task structure with high parallelism, and many tasks can be executed in concurrent. However, that of the Montage is inferior to LIGO, and Montage has tight tasks interdependence [141]. For these two real-world applications, we generate approximately 1, 000 tasks [69,112] and the edge communication between tasks are generated according to their CCR (the details can be seen in Section 5.5.2) [135,159]. This chapter mainly focuses on the parallel application with little data communication (in other word, the CCR is low). Thus, the CCR considered in this section varies between 0.05 and 0.4, in steps of 0.05. The workflow applications' deadlines are varied from 1.2 to 2 times the schedule length produced by HEFT algorithm, in steps of 0.1 [141].

7.5.3 The First Experimental Results

This section examines the performance sensitivity of the four algorithms relative to different *communication to computation cost ratios* (CCRs), the application deadlines for evaluation are set as 1.5 times schedule length produced by HEFT algorithm. The scheduling parallel application workflow of LIGO and Montage experimental results are shown in Figure 7.5 [141]. From

Figure 7.5(a), we observe that the average energy consumption of NEEWS significantly outperforms EATS by 14%, IC-PCP by 13.5%, and HEFT by 27.1%. In particular, the LIGO application with little data communication can save more energy consumption, such as $CCR = 0.05$. In fact, for such case, the presented NEEWS outperforms EATS by 17.8%, IC-PCP by 16.5%, and HEFT by 29.6%, respectively. This is mainly due to the fact that this workflow scheduling algorithm is a network energy-efficient approach, which can effectively deal withe high-speed fat-tree network routing chip (NWR) energy consumption. Moreover, the energy-aware EATS adopts DVFS to optimize processor resources and does not consider the idle network interconnection energy consumption.

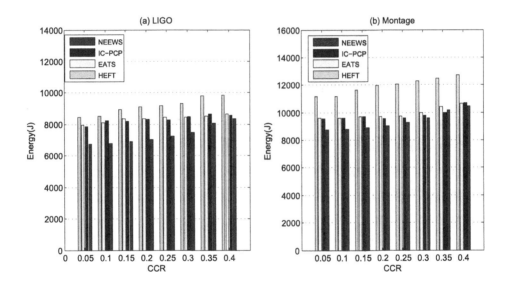

Figure 7.5 The results of varying CCR. (a) LIGO; (b) Montage.

From Figure 7.5(a), we can conclude that the LIGO energy consumption distinctly increases as the CCR increases for NEEWS and HEFT. This is due to the fact that the high CCR needs more data communication, and result in consuming more energy consumption. However, the energy consumptions of the IC-PCP and EATS are almost at the same level. Actually, as the CCR becomes 0.4, the performance difference among NEEWS,

EATS, and IC-PCP is not very distinct. These experimental phenomena demonstrate that the heuristic network energy-efficient task scheduling algorithm is not suitable for high CCR applications, such as data intensive applications.

The superiority of energy consumption savings also could be concluded from Figure 7.5(b), which provides the experimental results of Montage. In this experiment, the NEEWS algorithm outperforms EATS by 5.8%, IC-PCP by 4.8%, and HEFT by 31.7% in term of the average energy consumption. However, the energy savings of Montage are far less than that of LIGO. This is due the fact that Montage' tasks are strong interdependent. Therefore, there is difficulty in effectively sharing the high-speed interconnection network NWR. These experimental results also show that the heuristic network energy-efficient task scheduling algorithm is suitable for parallel applications with high parallelism and low data communication.

7.5.4 The Second Experimental Results

In this experiment, we vary the application deadline from 1.2 to 2 times the schedule length produced by HEFT algorithm, in steps of 0.1. The experimental results are shown in Figure 7.6. Here, the CCR is set as 0.1, which is used to denote the low communication parallel workflow application. We observe from Figure 7.6 that the energy consumption of NEEWS, EATS, and IC-PCP, slightly decreases as the deadline increases. In fact, for a high parallelism workflow application, such as LIGO (Figure 7.6(a)) and the NEEWS consumes almost the same energy. Actually, when the deadline is more than 1.7 times the schedule length produced by HEFT algorithm, the energy consumptions of this algorithm are almost same.

7.6 SUMMARY

This chapter first presents the computing nodes and high-speed interconnection networks of the systems, parallel application

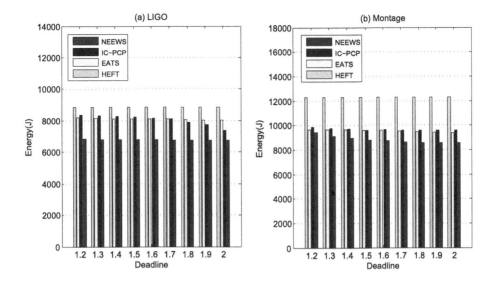

Figure 7.6   The results of varying deadline. (a) LIGO; (b) Montage

workflow DAG model, and basic concepts about energy consumption. Then, we present a heuristic network energy-efficient workflow task scheduling algorithm, which includes task level computing, task subdeadline initialization, task dynamic adjustment, and a data communication optimization scheme among tasks.

CHAPTER 8

Resource-Aware Duplication-Minimization Scheduling Algorithm

To satisfy the high-performance requirements of application executions, many kinds of task scheduling algorithms have been proposed. Among them, duplication-based scheduling algorithms achieve higher performance compared with others. However, because of their greedy feature, they duplicate parent tasks for each task as long as the finish time of the task can be advanced, which leads to a superfluous consumption of resource as well as energy. In addition, according to analysis, a large amount of duplications are unnecessary because slight delay of some uncritical tasks does not affect the overall makespan. In contrary, these redundant duplications occupy the resources, delay the execution of subsequent tasks, and might increase the makespan consequently. In this chapter, a duplication-efficient algorithm is introduced which can overcome the above drawbacks. The algorithm is to schedule tasks with the least redundant duplications. Besides, some optimizing strategies are introduced to search and remove redundancy for a schedule generated by the presented algorithm further.

DOI: 10.1201/b23006-8

8.1 INTRODUCTION

8.1.1 Definition of Task Scheduling

Task scheduling is to assign the tasks of an application to the processors of a target system based on different performance goals, such as minimizing the schedule length, satisfying the required reliability, or optimizing the energy consumption [124]. In the past decades, a lot of studies have been done on the task scheduling problem for constrained applications on heterogeneous computing systems.

A heterogeneous computing (HC) system usually refers to a distributed suite consisting of computing machines with different capabilities, which are interconnected by different high speed links and utilized to execute parallel applications [54, 94].

A constrained application usually refers to a large scale computing application which can be divided into a series of sub-tasks with dependencies. Task computation cost and inter-task communication cost of a constrained application are determined for an HC system via estimation and benchmarking techniques [31, 65, 158].

Finding a schedule with the minimal length for a given application is, in its general form, an NP-hard problem [55, 151]. Hence, many heuristic methods are proposed to obtain sub-optimal scheduling solutions [13, 14, 34, 59, 76, 92, 107, 108, 146]. In general, the task scheduling algorithms can be classified into a variety of categories, such as list scheduling algorithms, clustering algorithms, duplication-based algorithms, intelligent algorithms, and so on. Each type of algorithms has their respective advantages and disadvantages.

8.1.2   Introduction of Duplication-based Algorithms

Among all kinds of algorithms, the duplication-based algorithms can achieve a relatively good performance in terms of schedule length. The main idea of duplication-based algorithms is to reduce the communication overhead of a task with its parent tasks by duplicating its parent tasks locally, to achieve the goal of reducing the schedule length of the whole application.

A duplication-based algorithm is essentially a greedy algorithm, as each task is assigned to the processor which finishes the task earliest, and its parents are duplicated as long as the task finish time can be advanced. Whereas, the greedy strategy has some drawbacks as follows. First, under the greedy mechanism, most tasks in an application are executed multiple times, which leads to superfluous resource consumption and energy consumption. Second, duplicating some tasks just only advances the execution of their child tasks, but does not improve the whole schedule length. In other words, light delay of some uncritical tasks does not affect the overall schedule length, so the duplications of such uncritical tasks are redundant. Third, these redundant copies not only waste a huge amount of processor resources, but also delay the execution of the subsequent tasks, which increases the overall schedule length instead. In a schedule generated by the traditional duplication-based algorithms, there are a large amount of redundant duplications.

8.2   MODELS AND PRELIMINARIES

In Table 8.1, all the notations are summarized in order to improve the readability of this chapter.

TABLE 8.1 Notations used in this chapter

Notation	Description
P	the processor set in an HC system, $P = \{p_0, p_1, \ldots, p_{m-1}\}$
G	a DAG representing an application
T	the task set in G, $T = \{t_1, t_2, \ldots, t_n\}$
E	the communication cost matrix, $e_{ij} \in E$ denotes the communication cost between t_i and t_j
W	the computation cost matrix, $w_{i,j} \in W$ denotes the computation cost of t_i on p_j
$\overline{w_i}$	the average computation cost of task t_i
$pare_i(t_i)$	the immediate parent set of t_i
$pare_m(t_i)$	the mediate parent set of t_i
$child_i(t_i)$	the immediate parent set of t_i
$child_m(t_i)$	the mediate parent set of t_i
$ST(t_i, p_k)$	the actual start time of t_i on p_k
$FT(t_i, p_k)$	the actual finish time of t_i on p_k
$PBT(S)$	the processor busy time of a schedule S
$\Pi(t_i)$	the set of processors which are assigned to execute t_i
$rank_u(t_i)$	the upward rank of t_i
$EST(t_i, p_k)$	the earliest start time of t_i on p_k
$EFT(t_i, p_k)$	the earliest finish time of t_i on p_k
$DAT(t_i, p_k)$	the time that all required data arrives at p_k for the execution of t_i
$LFT(t_i, p_k)$	the latest finish time of t_i on p_k
$M_{i,k}$	the important immediate parent of t_i on p_k
H	the idle time slot set in a processor, which consists of a series of idle time slots
(t_i, p_k)	a copy of t_i on p_k
$C_f(t_i)$	the copy of t_i with the earliest finish time
$P_f(t_i)$	the processor where $C_f(t_i)$ is assigned
$T_{lc}(t_i, p_k)$	the local children set of (t_i, p_k)
$T_{oc}(t_i, p_k)$	the children copy set of t_i without local duplication of t_i
$T_{lp}(t_i, p_k)$	the local parent set of (t_i, p_k)
$T_{op}(t_i, p_k)$	the fixed copies of the off-processor parents of (t_i, p_k)

8.2.1 Computing System Model

A common distributed architecture is considered where a number of heterogeneous processors are mounted on the same network bus, which is shown in Figure 8.1. Let $P = \{p_0, p_1, \ldots, p_{m-1}\}$ be the set of m heterogeneous processors with different capacities, such as CPU frequency, random-access memory (RAM), etc. The computing cost of a processor executing tasks depends on how well the processor architecture matches the computation tasks. The execution time of a task when it is processed on a better-suited processor is shorter than that when it is processed on a unsuited processor. The best processor for one task may be the least suitable one for another task. A task can be assigned to any processor, and it should receive all required data from its predecessor tasks and send the required data to all of its successor tasks. Both of its predecessors and successors might be assigned to different processors. For example, in Figure 8.1, task t_1 is executed on p_1, and it sends data e_{12} and e_{13} to its two successor tasks t_2 and t_3 which are located on p_5 and p_4, respectively.

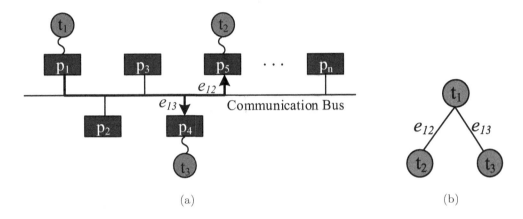

(a)　　　　　　　　　　　　　　(b)

Figure 8.1　Heterogeneous distributed system architecture.

8.2.2　Application Model

An application is represented by a directed acyclic graph (DAG) $G(T, E, W)$.

- $T = \{t_1, t_2, \ldots, t_n\}$ is the set of tasks in the application.

- $E = \{e_{ij}\}_{t_i \in T, t_j \in T}$ is the communication overhead matrix between tasks, where e_{ij} represents the communication time required to send data from t_i to t_j.

- $W = \{w_{i,k}\}_{t_i \in T, p_k \in P}$ is the computation cost matrix, where $w_{i,k}$ represents the computation time of task t_i on processor p_k.

A simple DAG is given in Figure 8.2 which consists of 13 tasks, and Table 8.2 lists the computation times of all tasks.

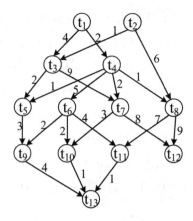

Figure 8.2 An example of DAG.

TABLE 8.2 WCETs of tasks on different processors

Tasks	p_0	p_1	p_2	p_3	$\overline{w_i}$
t_1	3	2	1	2	2
t_2	1	3	1	3	2
t_3	2	1	3	2	2
t_4	2	3	4	1	2.5
t_5	1	1	1	1	1
t_6	5	8	7	4	6
t_7	2	1	2	3	2
t_8	4	3	2	3	3
t_9	1	3	2	2	2
t_{10}	6	3	4	7	5
t_{11}	2	1	3	2	2
t_{12}	2	3	4	2	2.75
t_{13}	2	1	2	3	2

Next, some basic definitions are given about the DAG.

Definition 8.1 *In a DAG, t_i is termed as an* immediate parent *of t_j if $e_{ij} \in E$, that is, the execution of t_j depends on the output data of t_i. Correspondingly, t_j is termed as an* immediate child *of t_i. The immediate parent set of task t_i is denoted by $\mathrm{pare}_i(t_i)$, and the immediate child set of task t_i is denoted by $\mathrm{child}_i(t_i)$.*

Definition 8.2 *In a DAG, t_i is termed as a* mediate parent *of t_k, and t_k is termed as a* mediate child *of t_i, if $e_{ij} \in E$ and*

$e_{jk} \in E$. *The mediate child set and mediate parent set of task t_i are denoted by* $\text{child}_m(t_i)$ *and* $\text{pare}_m(t_i)$, *respectively.*

The *parent set* and *child set* of t_i, denoted by $pare(t_i)$ and $child(t_i)$, can be presented by $pare(t_i) = pare_i(t_i) \cup pare_m(t_i)$, and $child(t_i) = child_i(t_i) \cup child_m(t_i)$, respectively.

Example. For example, the immediate parent set and child set of task t_7 are $\{t_3, t_4\}$ and $\{t_{10}, t_{12}\}$, respectively. The mediate parent set and child set of task t_7 are $\{t_1, t_2\}$ and $\{t_{13}\}$, respectively.

Definition 8.3 *A task having no parent is called an* entry task, *and a task having no child is called an* exit task. *A DAG might have multiple entry tasks and multiple exit tasks.*

Example. For example, tasks t_1 and t_2 in Figure 8.2 are entry tasks, and tasks t_{12} and t_{13} are exit tasks.

In Table 8.2, the columns of p_0 to p_3 represent the computation cost of tasks on different processors. The average computation cost of task t_i is calculated as

$$\overline{w_i} = \frac{1}{n} \sum_{k=1}^{n} w_{i,k}.$$

The sample DAG is employed as an example throughout the following sections.

8.2.3 Performance Measures

To evaluate the performance of duplication-based algorithms, two important indicators are chosen, which are schedule length and processor busy time (PBT).

Because the original intention of task scheduling is the fastest execution of an application, the schedule length, or makespan, is one of the most important criteria undoubtedly.

Calculation of makespan. Given a schedule $S = \{(t_i, p_k, ST(t_i, p_k), FT(t_i, p_k)) | t_i \in T, p_k \in P\}$, where $ST(t_i, p_k)$

and $FT(t_i, p_k)$ represent the start time and finish time of t_i on p_k. Preemptive execution is not allowed, so

$$FT(t_i, p_k) = ST(t_i, p_k) + w_{i,k}.$$

The makespan is defined as

$$makespan = \max\{FT(t_i, p_k) \mid t_i \text{ is arbitrary exit task}\}. \quad (8.1)$$

In addition, due to resource awareness of the algorithm presented in this chapter, another criterion is utilized to measure the processor resource consumed by a schedule, which is defined as processor busy time (PBT). The PBT is the total time of processors executing tasks, and it also reflects the energy consumption of an application in some degree.

Calculation of PBT. Let $S = \{(t_i, p_k, ST(t_i, p_k), FT(t_i, p_k)) \mid t_i \in T, p_k \in P\}$ be a schedule generated by an algorithm, and then its processor busy time is calculated as

$$PBT(S) = \sum_{t_i \in T} \sum_{p_k \in \Pi(t_i)} (FT(t_i, p_k) - ST(t_i, p_k)), \quad (8.2)$$

where $\Pi(t_i)$ is the set of processors where t_i is executed.

Figure 8.3 shows a duplication-based schedule of the example DAG, whose processor busy time is 33 and makespan is 16.

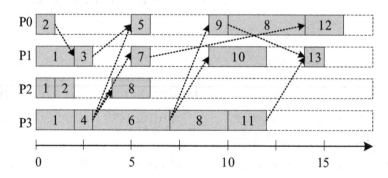

Figure 8.3 A duplication-based schedule of the example DAG.

8.3 RESOURCE-AWARE SCHEDULING ALGORITHM WITH DUPLICATION MINIMIZATION (RADMS)

The resource-aware duplication-minimization scheduling algorithm (RADMS) consists of three stages:

- Task prioritization stage: determine the priority of tasks;

- Task mapping stage: determine the mapping of tasks to processors;

- Redundancy deletion stage: search and delete redundant copies of tasks.

The algorithm framework of RADMS is given as Algorithm 14.

Input: The application G and the processor set P;
Output: The generated schedule S;
  1: Construct a task priority queue Γ; //Task Prioritization Stage
  2: $C \leftarrow \emptyset$;
  3: **while** Γ is not empty **do**
  4:     $t_i \leftarrow$ the first unscheduled task in Γ;
  5:     Determine the schedule of t_i; //Task Mapping Stage
  6:     **for** each $t_j \in T$ **do**
  7:         $child(t_j) \leftarrow child(t_j) - t_i$;
  8:         **if** $child(t_j) = \emptyset$ **then**
  9:            $C \leftarrow C \cup t_j$;
10:         **end if**
11:     **end for**
12:     **if** $C \neq \emptyset$ **then**
13:         **for** each $t_j \in C$ **do**
14:            Delete redundant copies of t_j; //Redundancy Deletion Stage
15:         **end for**
16:         $C \leftarrow \emptyset$;
17:     **end if**
18: **end while**

Algorithm 14: RADMS algorithm

8.3.1 Task Prioritization Stage

In RADMS, all tasks in a DAG are assigned with scheduling priorities based on upward ranking [146]. The task with the highest priority is scheduled first.

Calculation of upward ranks. The upward rank of a task t_i is recursively calculated by

$$rank_u(t_i) = \overline{w_i} + \max_{t_j \in child_i(t_i)} (e_{ij} + rank_u(t_j)), \qquad (8.3)$$

where $child_i(t_i)$ is the set of immediate children of task t_i. The rank value of exit task t_{exit} is

$$rank_u(t_{exit}) = \overline{w_{exit}}. \qquad (8.4)$$

The upward ranks of all tasks in the example DAG are listed in Table 8.3. From the example it can be concluded that the rank value of a task must be greater than all of its children, that means, a task should be scheduled earlier than all of its children.

TABLE 8.3　The upward ranks of tasks in the motivation application

Tasks	t_1	t_2	t_3	t_4	t_5	t_6	t_7	t_8	t_9	t_{10}	t_{11}	t_{12}	t_{13}
$rank_u$	30	27	24	23.5	12	16	13	15	8	8	5	2.75	2

8.3.2　Task Mapping Stage

The idea of task mapping is to allocate a task to the processor which completes it earliest. Let $EFT(t_i, p_k)$ be the earliest finish time of t_i when it is assigned to p_k. To determine the target processor for $t_i \in T$, it is necessary to calculate $EFT(t_i, p_k)$ for t_i on each processor $p_k \in P$.

Calculation of $EFT(t_i, p_k)$. In order to calculate $EFT(t_i, p_k)$, the finish time of all parents of t_i, which are scheduled before t_i, must be known in priori. Let t_p be one of the parents of t_i. It is assigned to p_c and its finish time is denoted by $FT(t_p, p_c)$. Since t_p can be assigned to multiple processors due to the duplication strategy, t_i tends to receive data from the copy whose data arrives earliest. Hence, the time that the data of t_p arrives at processor p_k for the execution of t_i, denoted as data arrival time (DAT), is calculated as

$$DAT(t_p, t_i, p_k) = \min_{p_c \in \Pi(t_p)} \{FT(t_p, p_c) + \overline{e_{p,i}}\}, \qquad (8.5)$$

where $\Pi(t_p)$ is the set of processors which have a copy of t_p, and $\overline{e_{p,i}}$ is the actual communication cost between t_p and t_i, which is calculated as

$$\overline{e_{p,i}} = \begin{cases} e_{p,i}, & p_k \neq p_c, \\ 0, & \text{otherwise.} \end{cases}$$

Once the schedule of all parents of t_i is determined, the final DAT for t_i is

$$DAT(t_i, p_k) = \max_{t_p \in pare_i(t_i)} \{DAT(t_p, t_i, p_k)\}, \qquad (8.6)$$

where $pare_i(t_i)$ is the immediate parent set of t_i, and the earliest finish time of t_i on p_k can be calculated by

$$EFT(t_i, p_k) = DAT(t_i, p_k) + w_{i,k}. \qquad (8.7)$$

From Eqs. (8.6) and (8.7), it can be concluded that $EFT(t_i, p_k)$ is mainly determined by the parent with the latest DAT. To distinguish different parents, an important definition is introduced as follows.

Definition 8.4 *Among all parents of a task t_i, the one whose data arrives latest is defined as the most important immediate parent (MIIP) of t_i.*

It is obvious that reducing the DAT of MIIP can minimize the EFT of a task. Algorithms 15 and 16 describe the process of task mapping and the detailed explains are given as follows.

1) First, calculate the earliest finish time of t_i on p_k when duplication strategy is not applied, which is denoted as $EFT'(t_i, p_k)$;

2) Second, determine the MIIP of t_i on p_k, and denote it as $M_{i,k}$. Notice that an entry task has not the MIIP;

3) Third, calculate the finish time of t_i on p_k if $M_{i,k}$ is duplicated on p_k before t_i, which is denoted as $EFT''(t_i, p_k)$;

4) Last, compare $EFT'(t_i, p_k)$ and $EFT''(t_i, p_k)$, and select the scheduling scheme which leads to a smaller EFT.

Input: Task t_i and a target processor p_k;
Output: The minimal $EFT(t_i, p_k)$ and the related schedule;
 1: Calculate $EFT'(t_i, p_k)$ when no duplication strategy is adopted;
 2: $M_{i,k} \leftarrow$ the MIIP of task t_i on p_k;
 3: **if** $M_{i,k}$ does not exist or is already scheduled on p_k **then**
 4:     $EFT(t_i, p_k) \leftarrow EFT'(t_i, p_k)$;
 5: **else**
 6:     Find the earliest available time slot on p_k for the execution of $M_{i,k}$;
 7:     Duplicate $M_{i,k}$ on p_k during the time slot and calculate $EFT''(t_i, p_k)$;
 8:     **if** $EFT''(t_i, p_k) < EFT'(t_i, p_k)$ **then**
 9:         $EFT(t_i, p_k) \leftarrow EFT''(t_i, p_k)$;
10:     **else**
11:         $EFT(t_i, p_k) \leftarrow EFT'(t_i, p_k)$;
12:     **end if**
13: **end if**

Algorithm 15: cal_EFT(t_i, p_k)

To duplicate the MIIP of task t_i on processor p_k, saying t_p, a suitable time slot should be exploited. Assume that H is the free time slot set on processor p_k, which consists of all free slots. A suitable time slot $[T_s, T_e]$ to duplicate t_p must satisfy

$$\max\{DAT(t_p, p_k), T_s\} + w_{p,k} \le T_e. \tag{8.8}$$

In all of the suitable time slots which satisfy Eq. (8.8), it selects the earliest one to duplicate t_p in order to minimize the EFT of t_i as much as possible. The example is given as Figure 8.4.

In Algorithm 15, the optimal schedule of t_i on each $p_k \in P$ can be determined as well as the $EFT(t_i, p_k)$ value. In Algorithm 16, they compare the schedule schemes of t_i being allocated to different processors, and determine the optimal processor p_k with the minimal $EFT(t_i, p_k)$. Till now, the schedule of t_i is determined.

Once the schedule of task t_i is determined, t_i is removed from the child set of all tasks. If there is a task whose child set becomes

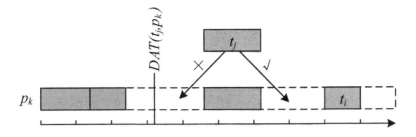

Figure 8.4 Determining the most-suitable time slot to duplicate t_j.

Input: Task t_i;
Output: The target p_k where t_i is assigned and the related schedule;
1: **for** each $p_k \in P$ **do**
2:    Call cal_EFT(t_i, p_k) and record $EFT(t_i, p_k)$;
3: **end for**
4: Schedule t_i on p_k with minimal $EFT(t_i, p_k)$;

Algorithm 16: Mapping of t_i

empty, RADMS turns to the redundancy deletion stage to search and delete the redundant copies.

8.3.3 Redundancy Deletion Stage

In the task mapping stage, a task is assigned to the processor which results in the minimal finish time. RADMS adopts a greedy strategy to duplicate the MIIP for each task as long as it can minimize the finish time of the task. Due to the greedy strategy, a task might have multiple copies in a schedule. In general, a task receives data from the parent copy whose data arrives earliest. However, if there is a parent copy executed on the same processor with the task, the task receives data from the local parent copy in priority; otherwise, it receives data from a parent copy executed on a different processor. In order to distinguish the two kinds of parent copies, a definition is introduced as follows.

Definition 8.5 *A task t_i is executed on p_k and t_c is one of its children. Assume that both of t_i and t_c have a copy on p_k, denoted as (t_c, p_k) and (t_i, p_k), respectively, and (t_i, p_k) is executed earlier* *than* (t_c, p_k), *then* (t_c, p_k) *receives data from* (t_i, p_k) *without communication. Here, the copy* (t_i, p_k) *is called a* local parent copy *of* (t_c, p_k), *and* (t_c, p_k) *is called a* local child copy *of* (t_i, p_k).

If the task copy of t_c *on* p_k *does not have a local parent copy of* t_i *and it has to receive data from a copy of* t_i *executed on a different processor, then the copy* $(t_i, p_r)(r \neq k)$ *is called an* off-processor parent copy *of* (t_c, p_k), *and* (t_c, p_k) *is called an* off-processor child copy *of* (t_i, p_r).

Due to the duplication strategy adopted in the second stage, a task might have several redundant copies on different processors.

Definition 8.6 *Assume that a task has multiple copies in a schedule. A copy is called as a redundant copy if deleting it does not affect the schedule of its children tasks.*

An example scenario is given to demonstrate how redundant copies are generated.

Example. A task t_i is mapped to p_k. Let t_c be a child of t_i which is mapped to a different processor p_r $(p_r \neq p_k)$, and t_i be the MIIP of t_c. To minimize the EFT of t_c on p_r, a duplicated copy of t_i is generated on p_r. In that case, the original copy of t_i on p_k might become a redundant one.

Deleting redundant copies from a schedule does not affect the overall makespan, but brings two benefits: decreasing resource consumption and releasing processor resources for subsequent tasks.

Now, the most important issue in this stage is when to search and how to determine the redundant copies.

When to search. In RADMS, redundant copies of a task are searched when its all children are scheduled. That is because only at this time a conclusion can be done whether a task copy is really redundant, by analyzing if deleting it affects the execution of its children tasks as well as the whole makespan.

How to determine. A task copy can be deleted only when other copies of the task can provide data required by all of its children. To describe the process in detail, some definitions are given firstly as follows.

Definition 8.7 *In a schedule, a task t_i has multiple copies. The copy with the earliest finish time is defined as the* fixed copy *of t_i, denoted by $C_f(t_i)$. The corresponding processor is defined as the* fixed processor *of t_i, denoted by $P_f(t_i)$.*

Because the fixed copy finishes earliest among all copies of a task, it provides data for those children without local parent. Hence, the only function of other copies is to provide data for their local children.

Let t_p be a parent of t_i. t_p and t_i have a copy on processors p_c and p_k, separately. Then the copy (t_p, p_c) can provide data for (t_i, p_k) if it satisfies

$$FT(t_p, p_c) + \overline{e_{p,i}} \le ST(t_i, p_k). \tag{8.9}$$

Algorithm 17 shows the framework of redundancy deletion stage.

Input: The set of tasks C whose redundant copies are to be deleted, the original schedule S;
Output: The schedule S' after removing the redundant copies;
  1: **for** all tasks $t_j \in C$ in nondecreasing order of $rank_u$ **do**
  2:    **if** t_j has multiple copies **then**
  3:      **for** each processor p_k that has a copy of t_j **do**
  4:        $S' \leftarrow$ delete the copy (t_j, p_k) from the schedule S;
  5:        **if** the dependencies between tasks in S' are satisfied **then**
  6:          $S \leftarrow S'$;
  7:        **else**
  8:          Undo the deletion operation of (t_j, p_k);
  9:          $S \leftarrow S$;
10:        **end if**
11:      **end for**
12:    **end if**
13: **end for**

Algorithm 17: Redundancy deletion stage

Let (t_i, p_k) represent a copy of t_i on p_k. The process of determining whether (t_i, p_k) is redundant are described as follows:

1) Delete (t_i, p_k) from the original schedule S.

  – If (t_i, p_k) is the fixed copy of t_i, the copy of t_i with the second earliest finish time is selected as the new fixed copy.

2) Determine whether all dependencies between tasks in the new schedule S' are satisfied.

  – If (t_i, p_k) is the original fixed copy, judge if the new fixed copy can provide data for the children without local parent of t_i. If true, (t_i, p_k) is redundant; otherwise, it is not.

  – If (t_i, p_k) is not the original fixed copy, judge if the fixed copy of t_i can provide data for the local children of (t_i, p_k). If true, (t_i, p_k) is redundant; otherwise, it is not.

8.3.4 A Scheduling Example

In order to demonstrate the process of RADMS, a schedule is given for the example DAG as follows. The priorities of tasks are calculated by Eq. (8.3), and the task scheduling order is determined according to the priorities, which is $\{t_1, t_2, t_3, t_4, t_6, t_8, t_7, t_5, t_9, t_{10}, t_{11}, t_{12}, t_{13}\}$.

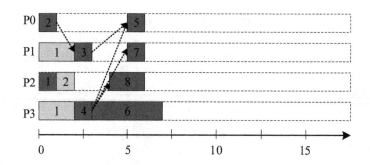

Figure 8.5 Schedule of tasks $t_1, t_2, t_3, t_4, t_6, t_8, t_7$, and t_5.

In Figure 8.5, tasks $t_1, t_2, t_3, t_4, t_6, t_8, t_7, t_5$ are scheduled in sequence and mapped to the processors which complete the tasks earliest. For example, to determine the schedule of t_3, the earliest finish times of t_3 on p_0, p_1, p_2, and p_3 under no replication policy are calculated firstly, which are given as follows.

- $EFT(t_3, p_0) = \max\{1 + 4, 1 + 0\} + 2 = 7$;

- $EFT(t_3, p_1) = \max\{1 + 4, 1 + 1\} + 1 = 6$;

- $EFT(t_3, p_2) = \max\{1 + 0, 1 + 1\} + 3 = 5$;

- $EFT(t_3, p_3) = \max\{1 + 4, 1 + 1\} + 2 = 6$.

When adopting the duplication strategy, the earliest finish times of t_3 on p_0, p_1, p_2, and p_3 are calculated as follows.

- $MIIP(t_3, p_0) = t_1$, duplicate t_1 on p_0, then $EFT(t_3, p_0) = 3 + 1 + 2 = 6$;

- $MIIP(t_3, p_1) = t_1$, duplicate t_1 on p_1, then $EFT(t_3, p_1) = \max\{2 + 0, 1 + 1\} + 1 = 3$;

- $MIIP(t_3, p_2) = t_2$, duplicate t_2 on p_2, then $EFT(t_3, p_2) = 1 + 1 + 3 = 5$;

- $MIIP(t_3, p_3) = t_1$, duplicate t_1 on p_3, then $EFT(t_3, p_3) = \max\{2 + 0, 1 + 1\} + 2 = 4$;

Comparing the results, t_3 is mapped to p_1 and its MIIP t_1 is duplicated on p_1 before t_3. Similarly, the schedule of t_4, t_6, t_8, t_7, and t_5 can be determined. The schedule is shown as Figure 8.5.

Till now, all child tasks of t_1, denoted as $child(t_1) = \{t_3, t_4, t_5, t_6, t_7, t_8\}$, have been scheduled. According to the algorithm, RADMS enters the redundancy deletion phase and the copy (t_1, p_2) is judged to be redundant and is removed from the schedule. The schedule after processing is shown as Figure 8.6.

Next, tasks t_9, t_{10}, t_{11}, and t_{12} are scheduled in sequence. After the schedule of t_{12} is determined, all children of t_2 have been scheduled, and the algorithm enters the redundancy deletion phase again. In this phase, no redundant copy is found.

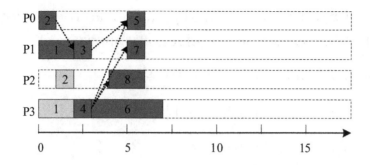

Figure 8.6 Schedule after deleting redundancy of t_1.

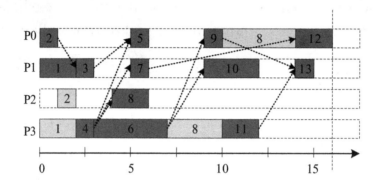

Figure 8.7 Schedule of tasks t_9, t_{10}, t_{11}, t_{12}, and t_{13}.

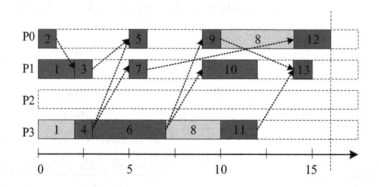

Figure 8.8 Schedule after deleting redundancy of t_2 and t_8.

After t_{13} is scheduled, the schedule is shown as Figure 8.7. Till now, the children of all tasks are scheduled, so all tasks should be judged if their copies are redundant. In the redundancy deletion stage, both of (t_8, p_2) and (t_2, p_2) are removed from the schedule, and the final schedule is given as Figure 8.8. It is apparently that the copeis generated by RADMS is three less than the schedule given in Figure 8.3.

8.4 DUPLICATION OPTIMIZING SCHEME

8.4.1 Analysis on Generation of Redundancy

The basic idea of all scheduling algorithms are to allocate a task to the processor which completes it earliest. To achieve this goal, duplication-based algorithms adopt the greedy mapping mechanism and duplicate the parent tasks for each task as long as doing this can bring forward the finish time of the task, which leads to a great number of duplicated copies generated.

In actual, the finish time of a task is mainly determined by its most important immediate parent, so it is unnecessary to complete all parents at the earliest time. Recall the schedule given in Figure 8.3, many tasks such as t_{13} and t_{10} can be shifted to a later time while not affecting the overall makespan. Through shifting, the duplicated copies of the parents of some tasks might become redundant since the time interval between tasks becomes longer and it might be long enough for the tasks to receive data from the fixed copy of their parents instead of the duplicated copy.

Moreover, the task t_8 has two children t_{11} and t_{12}. When determining the schedule of t_{11} and t_{12}, t_8 is duplicated on p_0 and p_3 to minimize the earliest finish time of t_{11} and t_{12}, respectively. Then, the fixed copy of t_8 on p_2 becomes redundant. Besides, (t_2, p_2) is a duplicated copy for the execution of (t_8, p_2), so it becomes redundant as well.

Based on above analysis, a schedule generated by any duplication-based algorithm can be optimized further. In this chapter, several optimizing methods are introduced to delete redundancy further.

8.4.2 Strategies of Redundancy Exploitation

Based on the analysis on the schedule process, it can break the strong dependencies between task copies in the original schedule by shifting or migrating tasks, to convert some task copies to redundant ones.

To guarantee the feasibility of an operation, the new generated schedule should satisfy the following conditions.

- For each copy (t_i, p_k), there must be at least one copy of each parent which can provide data for it.

- The fixed copy of t_i can provide data for all children which having no local copy of parent t_i.

- The schedule length of the new schedule should not exceed the old one.

In the following, three phases of deleting redundancies are introduced in detail.

8.4.2.1 Move Tasks to the LFT

The first strategy is to move tasks as late as possible to break strong dependencies between tasks.

Definition 8.8 *The* latest finish time *of a task t_i on p_k, denoted as* $\mathrm{LFT}(t_i, p_k)$, *is the latest time that the copy (t_i, p_k) can be moved to such that dependencies between t_i and its immediate children can be maintained.*

To guarantee the correctness of a schedule, a task copy cannot finish later than its latest finish time. Let $\{t_i, p_k, ST(t_i, p_k), FT(t_i, p_k)\}$ be an element of schedule S, $T_{lc}(t_i, p_k)$ be the local child set of (t_i, p_k) and $T_{oc}(t_i, p_k)$ be the children copies of t_i without local duplication of t_i. To calculate the latest finish time of t_i on p_k, two situations are considered as follows.

- If $p_k = P_f(t_i)$, (t_i, p_k) is the fixed copy of t_i and must provide data for all its off-processor children in $T_{oc}(t_i, p_k)$ and its own local children in $T_{lc}(t_i, p_k)$. Thus,

$$LFT(t_i, p_k) = \min\left\{ \min_{(t_c, p_k) \in T_{lc}(t_i, p_k)} ST(t_c, p_k), \right.$$

$$\left. \min_{(t_c, p_r) \in T_{oc}(t_i, p_k)} \big(ST(t_c, p_r) - e_{ic}\big)\right\}. \quad (8.10)$$

- If $p_k \neq P_f(t_i)$, (t_i, p_k) is a non-fixed copy. Then, (t_i, p_k) just needs to provide data for its local children in $T_{lc}(t_i, p_k)$, and its latest finish time is calculated by

$$LFT(t_i, p_k) = \min \left\{ \min_{(t_c, p_k) \in T_{lc}(t_i, p_k)} ST(t_c, p_k) \right\}. \qquad (8.11)$$

$LFT(t_i, p_k)$ is initialized as the makespan of the original schedule, denoted as L. After calculation, if $LFT(t_i, p_k) = L$ yet t_i is not an exit task, the copy (t_i, p_k) is determined as a redundant copy and deleted.

Once the latest finish times of all task copies are calculated, the task copies can be moved to finish as late as possible. By doing this, the strong dependencies between some task copies can be broken, and a certain of task copies become redundant. The pseudo-code of this strategy is shown in Algorithm 18.

Input: A schedule S generated by arbitrary duplication-based algorithm;
Output: The schedule S after deleting redundant copies;
1: $L \leftarrow$ the makespan of the input schedule S;
2: **for** each task t_i in nondecreasing order of $rank_u$ **do**
3:    **for** each processor p_k that has a copy of task t_i **do**
4:      Initialize $LFT(t_i, p_k)$ as L;
5:      **if** $p_k = P_f(t_i)$ **then**
6:        Calculate $LFT(t_i, p_k)$ by Eq. (8.10);
7:      **else**
8:        Calculate $LFT(t_i, p_k)$ by Eq. (8.11);
9:      **end if**
10:     **if** $LFT(t_i, p_k) = L$ and t_i is not an exit task **then**
11:       Delete (t_i, p_k) from S;
12:     **else**
13:       Move (t_i, p_k) to $LFT(t_i, p_k)$;
14:     **end if**
15:     Update the fixed copy of task t_i;
16:    **end for**
17: **end for**

Algorithm 18: Move tasks to the LET and remove redundancy

In the algorithm, the tasks are handled in nondecreasing order of ranks, that is to ensure that all children are processed before their parent tasks. The LFT values of all copies are initialized as L, the makespan of the input schedule. It ensures that our optimizing scheme does not deteriorate the performance of the original schedule.

Example. Consider a schedule shown in Figure 8.3. There are 13 tasks processed on four processors and the arrows in the schedule show important off-processor children dependencies. From the figure, it can be seen that tasks t_1, t_2, and t_8 have multiple copies, which are potential redundancy. According to Algorithm 18, tasks are traversed in the nondecreasing order of $rank_u$, that is $\{t_{13}, t_{12}, t_{11}, t_{10}, t_9, t_5, t_7, t_8, t_6, t_4, t_3, t_2, t_1\}$.

First, the tasks with single copy are moved to their LFT calculated by Eq. (8.10) or Eq. (8.11). For example, (t_{13}, p_1) is moved to 16, (t_{11}, p_3) is moved to 14, (t_{10}, p_1) is moved to 15. Considering t_8, since both of its children t_{12} and t_{11} have local parent on p_0 and p_3, the copy of (t_8, p_2) has neither off-processor children nor local children, so $LFT(t_8, p_2)$ is calculated as 16. Because t_8 is not an exit task and $L = 16$, (t_8, p_2) is removed from the schedule. The processing procedures of t_2 and t_1 are similar to t_8. (t_2, p_2) and (t_1, p_2) are deleted from the schedule too. The new schedule is shown as Figure 8.9.

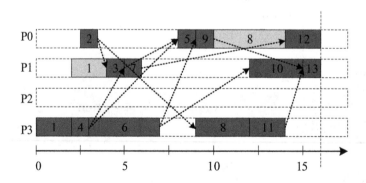

Figure 8.9 Schedule after tasks moving to the LFT and redundancy deletion.

8.4.2.2 Move Tasks to the EST

In the second strategy, the tasks having multiple copies are moved to start as early as possible. Doing so also can break strong dependencies between them and their children to exploit redundancy further. To formalize the start time of a copy, the following definition is given.

Definition 8.9 *The earliest start time of task t_i on processor p_k, denoted as $\mathrm{EST}(t_i, p_k)$, is the earliest time that t_i has received all required data from its parents and is ready for execution on p_k.*

After the copies of a task are moved earlier, the time intervals between them and their children get longer. A child copy which relies on the local parent could receive data from its off-processor parents, hence the local dependency is broken and the local parent could be deleted from the schedule.

To calculate the earliest start time of t_i on p_k, the data arrival times of all its parents must be known. Let t_p be a parent of task t_i. If (t_i, p_k) has a local copy of t_p, it receives data from the local copy (t_p, p_k) without communication; otherwise, it receives data from the fixed copy of t_p. The EST of t_i on p_k is calculated by:

$$EST(t_i, p_k) = \max \left\{ \max_{(t_p, p_k) \in T_{lp}(t_i, p_k)} FT(t_p, p_k), \right.$$

$$\left. \max_{(t_p, P_f(t_p)) \in T_{op}(t_i, p_k)} \left(FT(t_p, P_f(t_p)) + e_{pi} \right) \right\}, \quad (8.12)$$

where $T_{lp}(t_i, p_k)$ is the local parent set of (t_i, p_k), and $T_{op}(t_i, p_k)$ is the fixed copies of the off-processor parents of (t_i, p_k).

From Eq. (8.12), the finish time of t_i is determined by its parents, its parents must be processed before t_i. Moreover, the $EST(t_i, p_k)$ is not the actual time that t_i can be moved to, that is because the processor might be occupied by other tasks at that time. Therefore, it is necessary to find a proper slot which can accommodate the task, as introduced in Section 8.3.2.

The pseudo-code of the second strategy is described in Algorithm 19.

Example. Consider a schedule shown in Figure 8.9. Tasks are traversed in the nonincreasing order of $rank_u$, that is $\{t_1, t_2, t_3,$

Input: A schedule S after processing by the first strategy;
Output: The schedule S after shifting tasks and deleting redundancy;
1: **for** each task t_i in nonincreasing order of $rank_u$ **do**
2:    **if** t_i has no parent **then**
3:       **for** each processor p_k having a copy of t_i **do**
4:          $EST(t_i, p_k) \leftarrow 0$;
5:          Search a proper slot $[t_s, t_e]$ which can accommodate the task;
6:          Shift (t_i, p_k) to $\max\{EST(t_i, p_k), T_s\}$;
7:       **end for**
8:    **end if**
9:    **if** t_i has parents **then**
10:      **for** each processor p_k having a copy of t_i **do**
11:        Calculate $EST(t_i, p_k)$ by Eq. (8.12);
12:        Search a proper slot $[T_s, T_e]$ which can accommodate the task;
13:        Shift (t_i, p_k) to $\max\{EST(t_i, p_k), T_s\}$;
14:      **end for**
15:    **end if**
16:    Update the fixed copy of task t_i;
17: **end for**

Algorithm 19: Shift tasks to the EST and delete redundancy

---

t_4, t_6, t_8, t_7, t_5, t_9, t_{10}, t_{11}, t_{12}, t_{13}}. The first two tasks to be processed are t_1 and t_2. Both of them have no parent, so they are moved backward and start at time 0. Tasks t_3, t_4, and t_6 keep unchanged. Next, t_8 is considered and it has two parents t_2 and t_4, respectively. Since all parents of t_8 have been processed, the EST values of t_8 on p_0 and p_3 can be calculated, which are 4 and 7, respectively. In the example, t_8 on p_0 jumps over t_5 and t_9, starts at time 4 and ends at time 8. t_8 on p_3 is moved to start at time 7. Figure 8.10 shows the schedule at the end of this phase.

The second strategy aims at lengthening the time intervals between a task and its children, which makes preparation for the merging operation of next strategy.

8.4.2.3 Migrate Tasks among Processors

The previous two strategies are moving tasks forward or backward to break dependencies between tasks. The third strategy is to migrate tasks among processors.

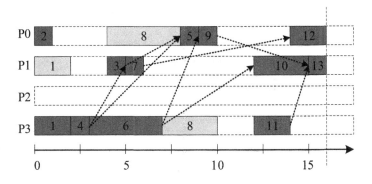

Figure 8.10 Schedule after tasks moving to the EST and redundancy deletion.

Thinking about the principle of duplication-based algorithms, the children of a task are allocated to different processors since there is no enough time in a processor to accommodate them. To finish as early as possible, they duplicate their MIIP greedily, which wastes a lot of resources. After the first two phases, a lot of redundant copies are deleted, so processors generate much idle time compared with the original schedule. Hence, a child task can be migrated to another processor where its brother is executed if the processor has enough idle time. By doing this, the duplicated MIIP of the child becomes redundant and can be removed. In the following, it gives a detailed introduction on the strategy.

The target tasks of this strategy are those tasks with multiple copies. Let t_i be a task with more than one copies. If the local child tasks depending on copy (t_i, p_k) can be migrated to another processor which has a copy of t_i, the copy (t_i, p_k) becomes redundant. To find out all redundant copies of t_i, all copies of t_i are grouped in pairs. For example, $\langle p_s, p_t \rangle$ denotes a pair of task copies of t_i on p_s and p_t, respectively. For each copy pair $\langle p_s, p_t \rangle$, it judges if (t_i, p_s) can be deleted by migrating its relied child tasks to p_t. Here, p_s is called a source processor and p_t is called a target processor. Each pair of t_i is assigned with a priority based on the execution time difference between two copies in the pair. The execution time difference of $\langle p_s, p_t \rangle$ is calculated as

$$D^i_{s,t} = w_{i,s} - w_{i,t}. \tag{8.13}$$

A pair with greater execution time difference is assigned with a higher priority.

For each pair $\langle p_s, p_t \rangle$ of t_i, they discuss if task t_i can be deleted from p_s under the help of copy (t_i, p_t). The judging process is described as follows.

- If (t_i, p_s) is the fixed copy of t_i which should provide data for its off-processor children, it cannot be deleted. If (t_i, p_s) is the fixed copy of t_i which has no off-processor children, it can be considered as a duplicated copy and the processing is shown as the next item.

- If (t_i, p_s) is a duplicated copy which has local children, they discuss the problem in the following situation. Let t_c be a local child of (t_i, p_s).

  – If (t_c, p_s) can receive data from another copy of t_i on a different processor, (t_c, p_s) is unnecessary to migrate;

  – Otherwise, judge if (t_c, p_s) can be migrated to the target processor p_t. The EST and LFT values of t_c on p_t are calculated, respectively. If t_c is migrated to p_t, it should be executed during time interval $[EST(t_c, p_t), LFT(t_c, p_t)]$ to satisfy the dependence. If there is an idle time slot $[T_s, T_e]$ on p_t which satisfies

$$\max\{T_s, EST(t_c, p_t)\} + w_{c,t} \leq \min\{T_e, LFT(t_c, p_t)\},$$
(8.14)

  then t_c is inserted to the slot; otherwise, (t_c, p_s) cannot be migrated to p_t.

  If all local children of (t_i, p_s) can either receive data from the off-processor parent, or be migrated to the target processor, (t_i, p_s) can be considered as a redundancy and deleted.

It is worthy mentioning, the tasks should be traversed in a nondecreasing order of $rank_u$. The algorithm is shown as Algorithm 20.

Input: A schedule S after processing by the first two strategies;
Output: The schedule S after task migration and redundancy deletion;
1: **for** each task t_i that has multiple copies in nondecreasing order of $rank_u$ **do**
2:     $Q \leftarrow \{\langle p_s, p_t \rangle | (t_i, p_s) \in S, (t_i, p_t) \in S\}$;
3:     **while** Q is not empty **do**
4:         $\langle p_s, p_t \rangle \leftarrow$ pop the first element in Q;
5:         $flag \leftarrow$ true;
6:         **if** (t_i, p_s) is the fixed copy which has off-processor children **then**
7:             **continue;**
8:         **else**
9:             **if** $T_{lc}(t_i, p_s) \neq \emptyset$ **then**
10:                 **for** each copy $(t_c, p_s) \in T_{lc}(t_i, p_s)$ **do**
11:                     **if** another copy of t_i can provide data for (t_c, p_s) **then**
12:                         **continue;**
13:                     **else**
14:                         Calculate $EST(t_c, p_t)$ and $LFT(t_c, p_t)$;
15:                         **if** a copy of t_c exists on p_t during $[EST(t_c, p_t), LFT(t_c, p_t)]$ **then**
16:                             Delete (t_c, p_s);
17:                             **continue;**
18:                         **else**
19:                             **if** an idle slot satisfying Eq. (8.14) can be found in p_t **then**
20:                                 Delete (t_c, p_s) and insert t_c into p_t;
21:                             **else**
22:                                 $flag \leftarrow$ false;
23:                                 **break;**
24:                             **end if**
25:                         **end if**
26:                     **end if**
27:                 **end for**
28:             **end if**
29:         **end if**
30:         **if** $flag$ =true **then**
31:             Delete (t_i, p_s) if it becomes redundant;
32:             $Q \leftarrow Q - \{\langle p_i, p_j \rangle | i = s \text{ or } j = s\}$;
33:         **end if**
34:     **end while**
35: **end for**

Algorithm 20: Migrating tasks among processors

Example. Consider a schedule shown in Figure 8.10. t_8 is the first considered task, and its pair queue is $Q = \{\langle p_0, p_3 \rangle, \langle p_3, p_0 \rangle\}$. The process of $\langle p_0, p_3 \rangle$ is as follows. First, (t_8, p_0) doesn't need to provide data for its off-processor children, and it has only one local child t_{12}. Since (t_8, p_3) cannot provide data for (t_{12}, p_0), they judge if t_{12} can be migrated to p_3. The EST and LFT values of t_{12} on p_3 are 14 and 16, respectively. The idle slot $[14, 16]$ on p_3 is available for t_{12}. Hence, migrating t_{12} to p_3 is feasible, and (t_8, p_0) becomes redundant and is deleted from the schedule. Figure 8.11 shows the schedule after Phase 3.

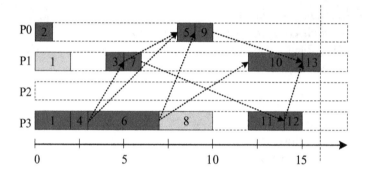

Figure 8.11   Schedule after tasks migration.

8.5   EXPERIMENTAL RESULTS AND ANALYSIS

Simulation experiments are adopted to validate the effectiveness and efficiency of RADMS and the three redundancy optimizing strategies.

8.5.1   Experimental Metrics

Two main performance measures are adopted in the experiments, which are makespan and resource consumption. Since resource consumption of applications varies with the number of tasks and has a large variation range, it is necessary to normalize the resource consumption firstly. Here, it defines the normalized resource consumption (NRC) as an indicator of measuring resource consumption:

$$NRC = \frac{PBT(S)}{PBT_{lower}}, \tag{8.15}$$

where $PBT(S)$ is the resource consumed by a schedule S, and PBT_{lower} represents the absolute lower bound on the resource consumed by an application. The PBT_{lower} value is calculated as

$$PBT_{lower} = \sum_{i=1}^{n} \min_{j=0}^{m-1} \{w_{i,j}\}. \tag{8.16}$$

Apart of NRC, the number of duplications in a schedule is adopted as another metric which aims at giving a clear picture on how many duplications are deleted by our algorithms.

8.5.2 Parameter Settings

In order to verify the performance of RADMS and the optimizing polices for different types of applications with different characteristics, the random DAGs are utilized. To generate different types of DAGs, three fundamental characteristics are considered:

- DAG size n: The number of tasks in the application DAG, which is to determine the application size.

- Communication to computation cost ratio CCR: The average communication cost divided by the average computation cost of the application DAG, which is to determine if the application is computation-intensive or communication-intensive.

- Parallelism factor λ: The number of levels of an application DAG is generated randomly using a uniform distribution with mean value of $\sqrt{n}/\lambda$ and rounded up to the nearest integer. The width is generated using a uniform distribution with mean value of $\lambda\sqrt{n}$ and rounded up to the nearest integer. A low λ leads to a DAG with a low parallelism degree.

In the random DAG experiments, the number of tasks is selected from the set $\{100, 200, 300, 400, 500\}$, and both λ and CCR are chosen from the set $\{0.2, 0.5, 1.0, 2.0, 5.0\}$. To generate a DAG with a given number of tasks, λ, and CCR, the number of levels is determined by the parallelism factor λ firstly, and then the number of tasks at each level is determined. Edges are generated only between the nodes in adjacent levels, obeying a 0–1 distribution. Each task is assigned with a computation cost from a given interval following a uniform distribution. To obtain the desired CCR for a graph, the communication cost is also randomly selected with a uniform distribution, whose mean depends on the product of CCR in $\{0.2, 0.5, 1.0, 2.0, 5.0\}$ and the average computation cost.

For each group of parameter setting, 50 DAGs are generated and scheduled to avoid scattering effects. The results are averaged over the 50 graphs.

8.5.3   Experimental Results and Analysis

In this part, the performance of the presented algorithm RADMS and three redundancy optimization strategies are verified by comparing with HLD.

8.5.3.1   Effect of Task Number

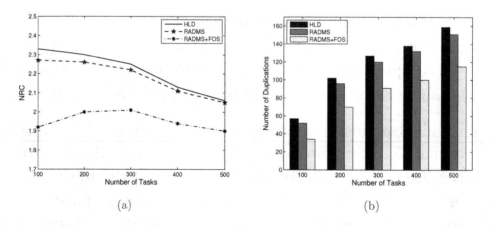

(a)   (b)

Figure 8.12   Effect of task number on performance.

The first group of experiments compare resource consumption and the number of duplications of the three algorithms with respect to various graph sizes, and the experimental results are shown in Figure 8.12. RADMS+FOS represents that RADMS is combined with three redundancy optimization strategies. It is known apparently that both RADMS and RADMS+FOS perform better than the HLD algorithm, and RADMS+FOS provides the smallest NRC on average. The average NRC value of RADMS on all generated graphs is reduced by 3% compared with the HLD algorithm. When combining with FOS, the ratio is up to 12%. From Figure 8.12(a), it can be noticed that the difference of NRC between RADMS+FOS and HLD decreases with the increasing number of tasks. The explanation is as follows. Under a fixed number of processors, as the number of tasks increases, a task is more prone to be assigned to the same processor with its parents, so less tasks are duplicated, which reduces the room of resource consumption improvement of the presented algorithms. From Figure 8.12(b), it can be seen that the number of duplications of three algorithms increases with the increasing number of tasks, and both of RADMS and FOS perform well in reducing the number of redundant copies.

8.5.3.2 Effect of Processor Number

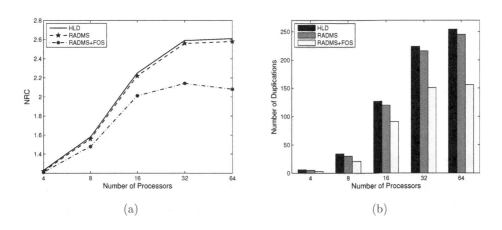

(a)   (b)

Figure 8.13   Effect of processor number on performance.

Figure 8.13 shows the experimental results with respect to different numbers of processors. Similarly, both RADMS and RADMS+FOS outperform the HLD algorithm. It can observe that the average NRC increases when the number of processors increases from 4 to 32, and keep still from 32 to 64. That is because duplication-based algorithms are prone to duplicate more tasks when there are enough idle processors, which leads to an increasing number of duplications, hence an increasing number of redundant copies. According to the experimental results, RADMS reduces resource consumption by 1.93% on average compared with HLD, while RADMS+FOS reduces by 11.25% on average, and the ratio is up to 20.24% with 64 processors.

8.5.3.3 *Effect of Parallelism Factor*

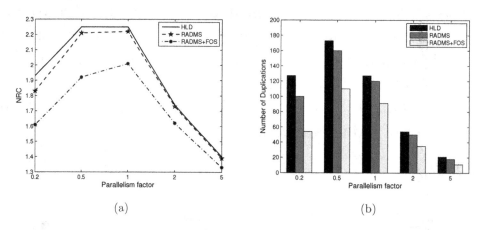

(a)   (b)

Figure 8.14   Effect of parallelism factor on performance.

In the third group of experiments, the average NRC values of the three algorithms are measured with various parallelism factor λ. Figure 8.14 shows that the average NRC values increases with the increasing λ at the beginning, which reaches the peak at $\lambda = 1.0$, and then decreases gradually from 1.0 to 5.0. Moreover, the NRC provided by RADMS is improved by 5.22%, 2.14%, 1.41%, 1.03%, and 0.61% compared with the HLD algorithm when λ is 0.2, 0.5, 1.0, 2.0, and 5.0, respectively. RADMS+FOS reduces the resource consumption by 16.32%, 14.88%, 10.66%, 7.38%, and 4.60%, respectively. The data show that the improvement becomes smaller as λ increases. Under a fixed number of tasks and processors, as λ gets smaller, the generated DAGs have a smaller parallelism. There are enough idle period on the processors to duplicate tasks, which is beneficial to RADMS and FOS. As λ increases, the feature becomes weaker, which deteriorates the performance improvement of RADMS and FOS.

8.5.3.4 Effect of CCR

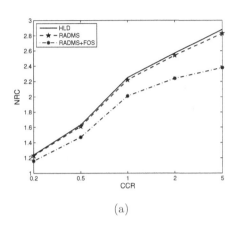 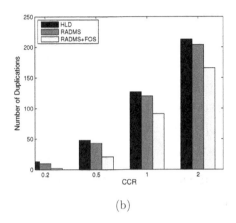

(a)   (b)

Figure 8.15   Effect of CCR on performance.

The last group of experiments aim at studying resource consumption of the three algorithms with respect to various CCR values. It is clear from Figure 8.15 that the average NRCs get greater with the increasing CCR. When CCR increases, the ratio of communication and communication cost increases, and the communication cost dominates the computation cost when CCR > 1. Tasks are assigned repeatedly to eliminate the communication between tasks, that is the reason that NRCs increase, hence the performance gaps between RADMS and HLD and between RADMS+FOS and HLD increase.

8.5.3.5 Makespan Improvement

Since makespan is an important measure to evaluate the performance of algorithms, they count the number of times that a schedule generated by RADMS has shorter, equal, and longer makespan compared with that generated by HLD, which are listed in Table 8.4. From the table, it can be seen that the percentage for RADMS to outperform HLD in terms of makespan

TABLE 8.4 A comparison of makespan for random DAGs

Parameter	Performance		
	Shorter	Equal	Longer
Number of tasks	86	151	13
Number of processors	93	138	19
Parallelism factor	76	151	23
CCR	126	94	30

is 38.1%, and the percentage that the schedules generated by the two algorithms have the same makespan is 53.4%. Overall, our algorithm improves the performance in terms of makespan compared with HLD. Nevertheless, the improvement is very small as existing duplication-based algorithms have already achieved strong performance in terms of makespan compared with listing scheduling algorithms.

8.6 SUMMARY

This chapter presents a duplication-efficient algorithm that can schedule tasks with the least redundant duplications. In addition, it also presents some optimizing strategies to search and remove redundancy for a schedule solution.

CHAPTER 9

Contention-Aware Reliability Efficient Scheduling

Energy efficiency and system reliability are the two main measurements in modern high-performance computing. The majority of previous recent studies have focused on realizing parallel task scheduling with low energy consumption or fast execution time. These methods were devised base on classic scheduling model. However, the contention model is being recognized by more and more researchers as a tool to develop accurate and efficient scheduling methods that are closer to the real environment. This chapter introduces a contention-aware reliability management with deadline and energy budget constraints (CARMEB) algorithm for priority constrained task scheduling in heterogeneous computing environments. CARMEB involves three phases, namely, task priority calculation, communication edge allocation, and slack reclaiming.

9.1 INTRODUCTION

Energy saving is a key problem to be solved in green computing. Modern high-performance processors, even idle ones, consume large amounts of energy. For instance, the Intel Core i7-975 3.33 GHz 1 MB L2, 8 MB L3 consumes 83 W when idle, and its peak power consumption reaches 210 W [84]. Moreover, the trend of

DOI: 10.1201/b23006-9

scaling transistors and operating voltages has markedly increased the susceptibility of processors to faults. For example, the soft-error failure rate at 16 nm is expected to be 100 times worse at 180 nm [96].

As a very popular and effective energy-saving technology, DVFS is extensively used in the algorithm design of energy-aware heterogeneous computing systems [61, 77, 78, 98, 170, 171]. Reliability is another primary performance metric in hetero-geneous systems [16, 21, 21, 85, 127]. Reliability maximization algorithm under an energy constraint [171], contention-aware, energy-efficient duplication (FastCEED) algorithm [123], reliable hierarchical reliability-driven scheduling (HRDS) algorithm [138], and reliable dynamic level scheduling (RDLS) algorithm [40] were developed to improve system reliability in HCS.

Researchers seldom consider communication contention in scheduling algorithms design of heterogeneous computing sys-tems. In practical application scenarios, this kind of communica-tion contention can be found everywhere in all computing systems [126, 181]. The present chapter utilizes a well-accepted communi-cation contention model with an energy budget on HCSs, aiming at maximizing reliability of priority-constrained tasks [172].

9.2 MODELS AND PRELIMINARIES

This section describes the target HCS, application model, power model, and reliability model utilized in this chapter.

9.2.1 Application Model

A target HCS with communication contention can be represented by $M_{HG} = HG = (PE, L)$, where PE comprises a set of DVFS-enabled processors, and L is the communication network with contention. This dedicated system has the following character-istics: (1) No time cost exists for local communication; and (2) communication is completed in a communication subsystem.

TABLE 9.1   Notations used in this chapter

Notation	Definition
HCS	The heterogeneous computing system
PE	A set of processing elements
V	A set of supply voltages
F	A set of supply frequencies
L_i	A link, transfers data between the processors and the switch
$w_{i,j}$	The task $t_i \in T$ executed on processor $pe_j \in PE$
$c_{i,j}$	The communication cost between node t_i and node t_j
$\bar{w}_i$	The average computational time of a task when executed on different processors
$EST(t_i, pe_j)$	The earliest execution start time of task t_i on processor pe_j
E_{ij}^L	The communication energy consumption from p_i to p_j
E_{total}	The total energy consumption of processors while performing all tasks in a task set
$EFT(t_i, pe_j)$	The earliest execution finish time of task t_i on processor pe_j
$R_i(f_i)$	The probability of the task t_i when it executes successfully
$t_f(t_i, pe_{src})$	The finish time of task t_i on processor pe_{src}
$t_{dr}(t_i, pe_k)$	The earliest time of node t_i can release on processor p_k
AFLs	Available frequency levels
DAG	Directed Acyclic Graph
DVFS	Dynamic Voltage Frequency Scaling
CCR	Communication to Computation Ratio
SLR	Scheduling Length Ratio
ECR	Energy Consumption Ratio
POF	Probability of Failure
RDLS-CA	The Reliable Dynamic Level Scheduling algorithm under the contention model
HRDS-CA	The Hierarchical Reliability-Driven Scheduling algorithm under the contention model
CARMEB	The contention-aware reliability management with energy budget algorithm
S_n	The total number of nodes of a special DAG

A parallel application that involves a set of precedence-constrained tasks can be described as a DAG, specifically a two-tuple DAG, $G = \langle T, A \rangle$ (the details can be seen in Section 1.3.2), where T is the parallel task set, and A comprises the edges that indicate the precedence constraints among tasks. Figure 9.1 shows a simple DAG. The notations employed in this chapter are summarized in Table 9.1.

9.2.2   Communication Contention Model

The communication system in the traditional model is free of contention. Every processor can send/receive messages to/from

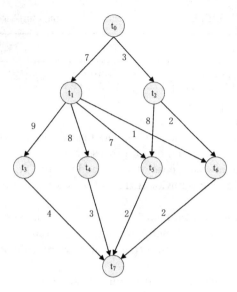

Figure 9.1 Simple DAG.

another processor at any time under this environment. However, this model type is unrealistic. The processors in this chapter are connected to a central switch as depicted in Figure 9.2, which shows that the target communication system model is a star network. The two-tuple on switches and links denotes their unit busy and idle energy consumptions [123]. Notably, all three links are of unit bandwidth. If remote communication occurs (e.g., t_i is assigned to pe_1 where its immediate parent is assigned to pe_0), then the edge between t_i and its immediate parent is scheduled on links L_0 and L_1 (route from pe_0 to pe_1).

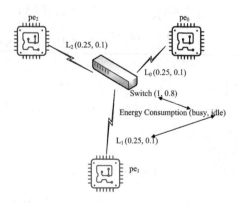

Figure 9.2 Link model.

9.2.3 Energy Model

As mentioned in Section 6.2.3, when all frequencies are formalized over the maximum frequency f_{max}, the energy consumption $E_i(f_i)$ of task t_i can be referred to Eq. 6.2.

In this chapter, contention appears among the links when communication occurs among the edges. And the energy consumption of the links is of unit bandwidth [123]. The communication energy consumption is derived as follows:

$$
\begin{aligned}
E_{i,j}^L &= E_i^s + e_{ij} \cdot f_{max} \\
&= E_i^s + e_{ij},
\end{aligned}
\tag{9.1}
$$

where E_{ij}^L is the communication energy consumption from p_i to p_j, E_i^s is the start communication cost of link L_i, and e_{ij} is the communication volume. The speed of the communication system in this chapter is at the maximum frequency. Hence, the total energy consumption E_{total} can be estimated while every task is performed in a task set by the following:

$$
E_{total} = \sum_{i=1}^{n} E_i(f_i) + \sum_{i=1}^{n} \sum_{j=1}^{n} E_{ij}^L + E_S,
\tag{9.2}
$$

where E_S is the switch energy consumption.

The processors in the HCS investigated in this chapter are DVFS-enabled and have different computation capabilities. Only the computational energy is considered for simplicity (i.e., the first item in Eq. (9.2)). Table 6.1 lists the voltage-frequency combinations of the heterogeneous processors.

9.2.4 Reliability Model

As introduced in Section 6.2.4, the system reliability R_{sys} indicates whether all of the n tasks in the task set are executed successfully without any transient faults. The expression can be described as follows:

$$
R_{sys} = \Pi_{i=1}^{n} R_i(f_i).
\tag{9.3}
$$

9.3 PRELIMINARIES

The topology of a communication network is modeled as a graph $NG = (T, PE, Q, \mathcal{L})$.

Definition 4 *(Edge Finish Time). Let $\mathcal{L} = \langle L_1, L_2, ..., L_l \rangle$ be the communication route of $e_{ij} \in E$, which sends data from pe_{src} to pe_{dst}. The finish time of e_{ij} is as follows:*

$$t_f(e_{ij}, pe_{src}, pe_{dst}) = \begin{cases} t_f(t_i, pe_{src}) & if \ pe_{src} = pe_{dst} \\ t_f(e_{ij}, L_l) & otherwise, \end{cases} \quad (9.4)$$

where $t_f(t_i, pe_{src})$ is the finish time of task t_i on processor pe_{src}. Thus, $t_f(e_{ij}, pe_{src}, pe_{dst})$ is the finish time of e_{ij} on the last link of the route, which is referred to as L_l. Unless pe_{src} and pe_{dst} are the same processor, no communication occurs. In particular, the communication is local.

In the contention-aware model, the earliest time of node t_i that can be released on processor p_k is defined as the data-ready time t_{dr}, which can be expressed as follows:

$$t_{dr}(t_i, pe_k) = \max_{e_{ij} \in Q, t_i \in prev(t_j)} \left\{ t_f(e_{ij}, pe(t_i), pe_k) \right\}, \quad (9.5)$$

where $pe(t_i)$ refers to the processor allocation of node t_i, $pe(t_i) \in PE$.

Definition 5 *The immediate neighboring frequency f_{in} of f_{ee} (energy efficient frequency) on processor pe_j is the selected frequency to save more energy.*

Given that $f_{ee} = \sqrt[3]{\Phi_{ind}/(m-1)\zeta}$ and because m is a constant (with a typical value of $m=3$), f_{ee} depends only on Φ_{ind} [173].

9.3.1 Task Priority

A feasible and efficient prior order, introduced in Eq. (6.5), is established in this chapter to comply with the precedence constraint of parallel tasks.

Table 9.2 lists the computation costs for each node on different processors. The last column of Table 9.2 indicates the corresponding *URank* value for each node in Figure 9.1.

TABLE 9.2 Computation costs on different processors

Tasks	pe_0	pe_1	pe_2	*URank*
t_0	5	7	8	45.67
t_1	6	7	3	33.00
t_2	7	10	5	41.00
t_3	6	8	5	17.67
t_4	8	5	7	17.67
t_5	8	11	6	17.00
t_6	7	5	9	16.33
t_7	6	9	7	7.33

9.3.2 Problem Description

The current chapter problem is selecting an appropriate frequency for each task (in a specific priority order) given (a) a parallel task application, (b) processors in an HCS that support the DVFS technique, and (c) contention in the communication system among these processors. The system reliability should also be maximized simultaneously without exceeding the given deadline D^* and energy budget E^* constraints. This frequency should then be assigned to an available processor. The scheduling problem for parallel tasks in HCSs with communication contention is formalized as follows:

$$\text{Maximize:} \quad R_{sys},$$

$$\text{subject to:} \quad D_{total} \leq D^*, \tag{9.6}$$

$$E_{total} \leq E^*. \tag{9.7}$$

9.3.3 Motivational Example

A motivational example is presented in this section. The scheduling of a task graph in Figure 9.1 is illustrated in Figure 9.3. Figure 9.3(a) shows the task scheduling in the computing systems free of

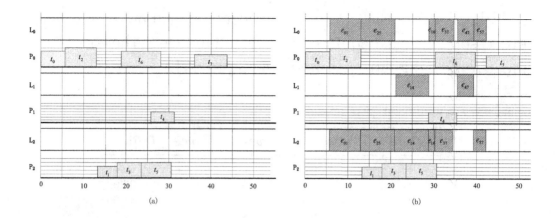

Figure 9.3 Scheduling of task graph in Figure 9.1. (a) schedule without contention; (b) schedule under CARMEB with contention.

contention. Figure 9.3(b) captures the task scheduling in a communication contention system. a_{01} is scheduled before the release of t_1 as shown in Figure 9.3(b). As its immediate parent τ_0 is assigned to pe_0, t_1 is scheduled to pe_2. Remote communication then occurs. The data should be transferred from pe_0 to pe_2. The route path is $pe_0 \rightarrow L_0 \rightarrow switch \rightarrow L_2 \rightarrow pe_2$. Thus, a_{01} is scheduled in L_0 and L_2. t_2 is ready to release once it receives these data. t_6 waits for the data from its immediate parent t_1 before its release. However, a_{16} is delayed in schedule, because a_{14} has occupied the communication link a_{12}. The deadline and energy budget for Figure 9.1 are 51 and 95, respectively. Under this scheme, the total energy consumption of the processors is 89.086, with a probability of failure $(1 - R_{sys})$ is 3.42e-07. The makespan values are 44.58 and 49.96 in Figures. 9.3(a) and 9.3(b), respectively.

9.4 CONTENTION-AWARE RELIABILITY MANAGEMENT SCHEME

Reliability is the most important measurement in heterogeneous computing. Therefore, the objective is to maximize the system reliability of task scheduling while guaranteeing a total energy consumption of processors that does not exceed an energy budget E^* and a makespan that does not exceed a deadline D^*.

This section devises a novel contention-aware reliability management with energy budget (CARMEB) algorithm based on a communication contention system. CARMEB performs each task assignment in a priority order, selects an appropriate execution frequency to attain higher reliability, and determines routes for the communications sent to this task. The pseudo code of CARMEB is described in Algorithm 21.

Input: A DAG $G =< T, Q >$, a M_{HG} and a set PE of DVFS available processors available processors.

Output: A schedule S of G onto M_{HG}.

1. Compute $URank$ of $t_i \in T$ by traversing the graph from the exit node;
2. Compute energy-efficient frequency for each node in set T;
3. Sort the tasks in a non-increasing order by $URank(t_i)$ value and establish a priority queue $Queue_{URank}$ for the sorted tasks;
4. **while** the priority queue $Queue_{URank}$ is not empty **do**
5. $t_i \leftarrow$ the head node in $Queue_{URank}$;
6.   **for** each processor $pe_j \in P$ **do**
7.   Calculate the summation of $t_{dr}(t_i, pe_j)$ and c_{ij};
8.   **for** $\forall f_{j,k} \in F$ **do**
9.   Find the immediate neighboring frequency f_{in} of f_{ee} on processor pe_j for task t_i, mark the one which has the earliest finish time;
10.   **if** $D_{total} \leq D^* \wedge E_{total} \leq E^*$ **then**
11. $f_{j,k} \leftarrow f_{in}$;
12.   **else**
13. $f_{j,k} \leftarrow f_{max}$;
14.   **end**
15.   **end**
16.   **end**
17.   Assign the marked frequency f_{in} to task t_i on the marked processor;
18.   Compute the reliability of task t_i using Eq. 6.3;
19.   Allocate e_{pi} to $L_{pe(t_p)}$ and $L_{pe(t_i)}$ when they are assigned to different processors // t_p is the immediate parent of t_i;
20.   Compute the energy consumption for task t_i using Eq. (6.2);
21.   Delete the head node t_i in $Queue_{URank}$;
22. **end**
23. $t_i \leftarrow t_{exit}$;
24. **while** $t_i \neq t_{entry}$ **do**
25.   Reclaim the previous scheduled tasks from t_{exit} upward to t_{entry}, adjust the candidate task in each slack with f_{ee};
26.   Update the reliability and energy consumption for the candidate task t_i;
27.   Update the candidate task;
28. **end**

Algorithm 21: CARMEB

The said algorithm shows that Steps 1 and 3 establish a priority queue for a DAG, such that the subsequent scheduling conforms to the predecessor constraint. Step 2 calculates the energy-efficient frequency that satisfies the energy budget condition. The schedule is prepared in Steps 4 to 22. Steps 8 to 15 establish the inner loop, which determines an efficient frequency for each candidate task on a particular processor without exceeding the deadline and energy budget. Step 17 assigns the most suitable frequency and processor for each candidate task. If remote communication is required, then edge scheduling begins from the processor where the scheduled candidate task is located. Steps 24 to 28 reclaim the slack in the prepared scheduling. The task order is upward from the exit task of the priority queue. When a slack occurs, the nearest task will update its execution frequency with the most efficient one. The time complexity of CARMEB is $O(|T| \cdot |PE| \cdot \log |F|)$, where $|T|$ is the DAG size, $|PE|$ is the processor number, and $\log |F|$ is the maximum level supported by the processor.

9.5 EXPERIMENTS

The presented algorithm is assessed in this section. A brief introduction of the three most popular existing algorithms (i.e., FastCEED, RDLS, and HDLS) is presented before making comparisons.

FastCEED [123] refers to the contention-aware, energy efficient, duplication heuristic algorithm based on mixed integer programming (MIP) formulation for parallel tasks scheduling in heterogeneous systems. The duplication idea is employed during task scheduling to minimize the makespan, total energy consumption, and tardiness of tasks in network resources. The communication energy consumption and latency are decreased significantly with the help of duplicating the most important parent task, which is obtained under the MIP method. The FastCEED shows an effect of energy efficiency and performance.

The HRDS was proposed by Tang et al. [138], which attempts to improve system reliability and decrease the scheduling length. HRDS includes local and global level scheduling. The parallel tasks of an application with precedence constraint run on heterogeneous processors of a virtual node under the local level scheduling strategy. By contrast, the independent applications are scheduled on the grid under the global level scheduling strategy.

The RDLS [40] algorithm was designed to consider the resource reliability in the HCS while minimizing the makespan of an application during the tasks scheduling. The DL of a task-processor pair plays an important part in RLDS, which involves four items. The first three items are used to promote the most suitable resources to minimize the makespan, whereas the last item is employed to improve the resource with higher reliability to enhance the application reliability. The task-processor pair with largest dynamic level is selected while making the decision for the candidate task mapping to a proper processor.

Comparing schedule algorithms under different models is generally unnecessary. Thus, the two aforementioned algorithms (i.e., RDLS and HRDS) are simulated with contention for comparison. In particular, the task scheduling in RDLS and HRDS maintains the same processor assignment in the same order, but the edges are scheduled for contention. These schemes are called RDLS-CS and HRDS-CS, respectively, for distinction.

Experiments were performed to analyze many aspects of the developed parallel task scheduling with contention-aware reliability management. Our experiments were conducted with a workstation equipped with an Intel Core i5-6400 quad-core CPU, 8 GB DRAM, and 2 TB disk. The operating system was Windows 7 (64 bit). As previously mentioned, the occurrence of transient faults follows a Poisson distribution. The parameter configurations utilized in this chapter were $d = 3$, $m = 3$, and $\lambda_0 = 10^{-9}$.

Parallel target systems were adopted in this chapter. The processor number was selected as 3, 6, 9, 15, 25, and 50. The switch employed in this chapter only supports half-duplex communication (i.e., only one communication in either direction was permitted at a time). The communication-to-computation ratio (CCR) is a highly important measure of a DAG. The CCR is defined as the summation of the entire edge weight over the entire node weight in a DAG (i.e., $CCR = \sum_{a \in A} w(a)/\sum_{t \in T} c(t)$).

The following subsections discuss the performance metrics, as well as randomly generated and real-world applications of DAGs, to evaluate the effect of the presented algorithm.

9.5.1 Performance Metrics

The performance of the presented algorithm is assessed in this section. Several important metrics must be introduced before the experimental results are evaluated. Randomly generated and real-world application DAGs are employed to verify the effects of presented algorithm and existing algorithms.

9.5.1.1 Scheduling Length Ratio (SLR)

The scheduling length, which is also called the makespan, is determined by the finish time of task t_{exit}. In particular, makespan = $t_{f(t_{exit})}$. The makespan is one of the key metrics for task scheduling. The makespan value becomes large when the DAG increases in size. The makespan should always be normalized. This metric is measured by the SLR, which is defined as follows:

$$\text{SLR} = \frac{\text{makespan}}{\sum_{t_i \in CP} \min_{pe_j \in PE}\{w_{i,j}\}}, \tag{9.8}$$

where CP denotes the task nodes located in the critical path of the DAG.

9.5.1.2 Energy Consumption Ratio (ECR)

The ECR for a given DAG is defined as Eq. (6.8).

9.5.1.3 POF

The POF indicates the probable execution fault of all tasks in a DAG that are mapped to specific processors with the allocated frequency under a particular strategy. The POF can be expressed as Eq. 6.9.

9.5.2 Randomly Generated DAG

Without loss of generality, a random DAG generated with a specific probability for an edge between any two nodes in the graph is considered in this chapter. Different characteristics are considered to capture this type of weighted-application DAG. The main parameters are listed as follows:

- DAG size: The number of task nodes in a DAG (ranges from 50 to 300).

- CCR: CCR=0.5, 1.0, 2.0, 5, 10.

- Number of processors: The processor number ranges from 3 to 20.

- Computation cost: The computation cost of each task varies uniformly from 5 to 50.

- Average in/out degree: The average in/out degree of each node ranges from 3 to 10.

- Heterogeneity factor (HF): HF is relative to the computation cost of each task, which varies in the range of 5 to 50.

The DAG graphs utilized in our experiments were generated through a combination of these parameters. The communication weight between two task nodes is created according to the CCRs. The probability of each communication edge follows a normal distribution.

9.5.3   Effect of Random Applications

These random applications are aimed at realizing better comparisons among our presented algorithms and three widely known algorithms (i.e., RDLS-CS, HRDS-CS, and FastCEED).

Figures 9.4 to 9.7 show that each datum in the charts, which denotes the average value of the result, is obtained after these algorithms were run for 100 times for a specific set of configuration parameters. Figures 9.4 to 9.6 show that the SLR, ECR, and POF significantly increase as CCR increases, especially when CCR is large. When CCR becomes equal to 0.5 (i.e., CCR is a computational intensive application), SLR, ECR, and POF are considerably less than those for CCR = 10. Figures. 9.4 to 9.6 also show that the average SLRs are extremely close to those for HRDS-CS and FastCEED. These results are due to ECR and POF being the two primary metrics in this chapter, whereas CARMEB only ensures that the SLR is modest. A light SLR must be sacrificed in several cases to guarantee a better POF. As expected, CARMEB outperforms the other three algorithms as task number increases for all CCRs. This performance improvement can be attributed to the intelligent strategy in the processor selection and frequency assignment of CARMEB for each task in a DAG. The scheduling of the communication edge is also considered on a parallel computing system with communication contention.

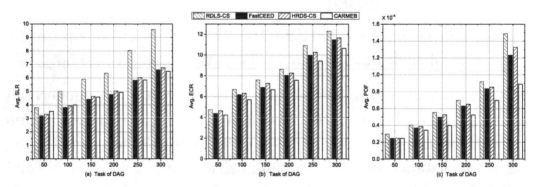

Figure 9.4   Effect of varying task number for CCR = 0.5.

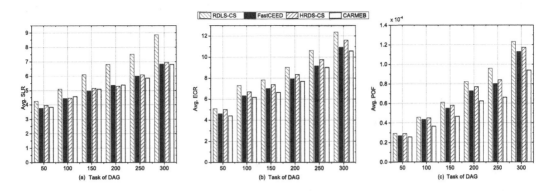

Figure 9.5 Effect of varying task number for CCR = 1.0.

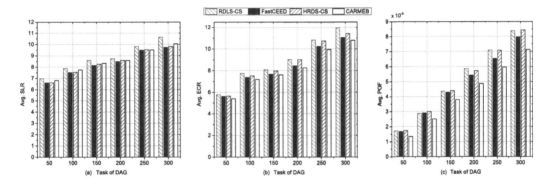

Figure 9.6 Effect of varying task for CCR = 10.

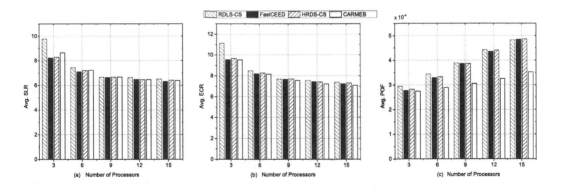

Figure 9.7 Effect of varying task number for CCR = 5 and DAGsize = 100.

Figure 9.6 illustrates the comparisons of the four algorithms for a large CCR. At CCR = 10, a communication intensive application can be observed. Unlike that of CCR = 1.0, the metrics in Figure 9.6 (i.e., SLR), are significantly increased. The reason for this condition is that the communication between two nodes takes much more time when the CCR is large. The said figure also shows that CARMEB surpasses RDLS-CS, HRDS-CS, and FastCEED by 8.59%, 6.36%, and 2.48% in terms of SLR, respectively. CARMEB significantly outperforms RDLS-CS, HRDS-CS, and FastCEED by 18.48%, 18.72%, and 12.58% in terms of the average POF, respectively. For the average value of SLR, CARMEB performs −3.15%, −1.26%, and 2.07% better than RDLS-CS, HRDS-CS, and FastCEED, respectively. These findings suggest that CARMEB suitably performs at a high CCR.

Figure 9.7 shows the performance of the four algorithms with varying task numbers for CCR = 5.0 and DAG size = 100. As the processor number increases, the SLR and ECR decrease consistently for the four algorithms, whereas the POF increases significantly. Among the four algorithms, CARMEB surpasses RDLS-CS, HRDS-CS, and FastCEED in terms of ECR and POF. Increasing the processor number does not always improve the SLR and ECR as shown in the case of 12 processors in Figure 9.7. The bottleneck in this scenario is not caused by the processor number but by the communication data from the parents of the candidate task in the scheduling.

9.5.4 Real-World Application DAG

Besides a randomly generated DAG, three real-world applications (i.e., LU decomposition [158], fast Fourier transform (FFT) [158], and molecular dynamics code [72]), are considered to obtain a comprehensive evaluation of the presented algorithm. The following subsections present that the experimental outputs denote the average value of the metrics after the four algorithms are run

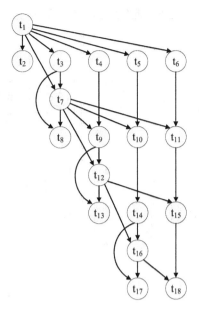

Figure 9.8 LU-decomposition task graph.

for 100 times with specific configurations. The metrics used for the comparison are SLR, ECR, and POF.

9.5.4.1 LU Decomposition

The structure of LU decomposition is shown in Figure 9.8. LU decomposition is widely utilized because of its capability to solve mathematical equations. The parameter configuration of LU graphs employed in this chapter is listed in Table 9.3. Let S_n be the DAG size of LU, wherein S_n satisfies the expression, $S_n = (n^2 + 3n)/2, (n \geq 3)$. S_n increases rapidly when n increases. As n increases to 30, the DAG size of LU reaches 495. The processor number varies in the range of 3, 6, 9, 12, 15, and 25. The CCR ranges from 1 to 10 with one-step increments.

TABLE 9.3 Parameter configuration for the LU task graphs

Parameter	Possible values
CCR	0.5, 1, 2, 3, 4, ..., 10
Number of processors	3, 6, 9, 12, 15, 25
Size of matrices	5, 6, 7, ..., 30, 31, 32

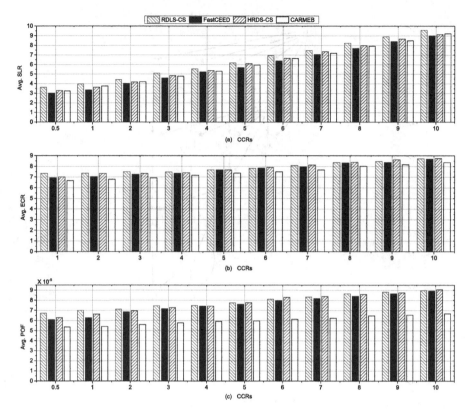

Figure 9.9 Effect of varying CCR the LU decomposition task graph.

The LU experimental results are shown in Figure 9.9. CARMEB surpasses RDLS-CS, HRDS-CS, and FastCEED for all CCRs in energy and reliability management. As CCR increases, the average values of ECR and POF for the four algorithms increase. For ECR, CARMEB performs 12.33%, 11.68%, and 9.89% better than RDLS-CS, HRDS-CS, and FastCEED, respectively. For the average POF value, the improvement of CARMEB is 31.00%, 29.47%, and 26.54% more than those of than DLS-CS, HRDS-CS, and FastCEED, respectively. For the average SLR, CARMEB performs 5.13%, 8.39%, and −3.37% better than DLS-CS, HRDS-CS, and FastCEED, respectively.

9.5.4.2 Fast Fourier Transform

The structure of FFT is shown in Figure 9.10. FFT is widely used in many fields, including science, mathematics, and engineering. Figure 9.10 illustrates a FFT with four points. It involves two parts. The first part is the area above the dashed lines. Tasks

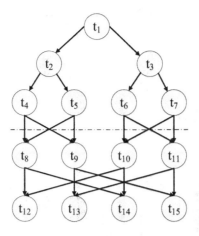

Figure 9.10 FFT with four points.

in this part recursively call tasks that are above them. The ones below the dashed lines are the tasks that apply the butterfly operation.

According to the characteristics of FFT, the DAG size increases quickly as the point varies. Let Sn be the FFT DAG size, where n is the point value. Thus, $S_n = 2^{n/2}$, $(n\%4 = 0)$; an exit task is added to each FFT graph. The parameter configuration utilized in our experiment is listed in Table 9.4. The DAG size reaches 512 when the point value becomes equal to 10. Figure 9.11 presents the average values of the experimental outputs, including the average values of the SLR, ECR, and POF, which are based on the combinations of the parameters in Table 9.4. As shown in Figures 9.11(b) and 9.11(c), the ECR and POF increase slightly as CCR increases. The ECR is notably large for all CCRs. In addition, CARMEB exhibits a more competitive ECR and POF than the other three algorithms.

TABLE 9.4 Parameter configuration for the FFT task graphs

Parameter	Possible values
CCR	0.5, 1, 2, 3, ..., 10
Number of processors	3, 6, 9, 15, 25
Size	4, 8, 12, 16, 20, 24

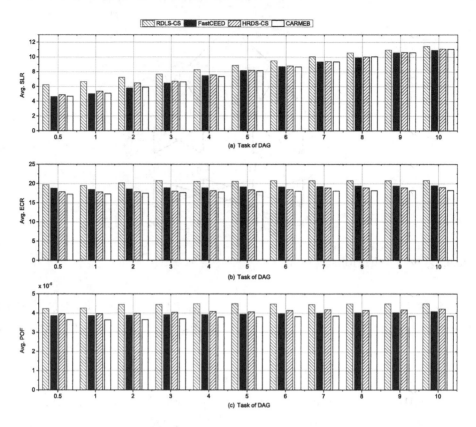

Figure 9.11 Effect of varying CCR for the FFT task graph.

Therefore, CARMEB surpasses RDLS-CS, HRDS-CS, and FastCEED for all CCRs in terms of ECR and POF. Specifically, the overall performance improvement of CARMEB for FFT is 17.59%, 10.90%, and 7.84% more than those of RDLS-CS, HRDS-CS, and FastCEED in terms of ECR, respectively. For the average POF value, CARMEB performs 29.68%, 22.02%, and 18.72% better than RDLS-CS, HRDS-CS, and FastCEED, respectively. For the average SLR value, CARMEB performs 8.51%, 1.38%, and −0.45% better than RDLS-CS, HRDS-CS, and FastCEED, respectively.

9.5.4.3 Molecular Dynamic Code

As shown in Figure 6.9, the DAG extracted from the molecular dynamics code presented in [72] is employed to evaluate the performance of the scheduling algorithms. This molecular dynamic

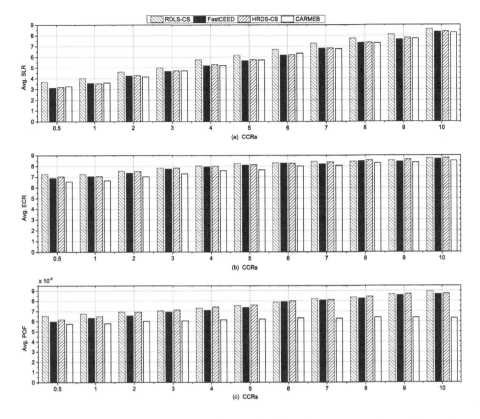

Figure 9.12 Effect of varying CCR for the molecular dynamics code task graph.

code graph is selected because of its irregularity. Given that its structure is known and that the task numbers are fixed, the experiment parameters are adjusted only within the range of the CCR, which ranges from 1 to 10 with increments of 1. Three processors are chosen in this experiment.

As observed in Figures 9.12(b) and 9.12(c), CARMEB outperforms the other three algorithms for all CCRs in terms of ECR and POF. Both the ECR and POF increase as CCR increases. On average, the overall performance of CARMEB is 5.56%, 4.99%, and 3.68% more than that of RDLS-CS, HRDS-CS, and FastCEED, respectively. For the average POF value, CARMEB performs 24.34%, 23.61%, and 20.59% better than RDLS-CS, HRDS-CS, and FastCEED, respectively.

9.6 SUMMARY

This chapter presents a contention-aware reliability management algorithm called CARMEB for priority-constrained tasks in heterogeneous computing environments. CARMEB algorithm can effectively solve the task scheduling problem of parallel workflows under communication contention in real heterogeneous computing environment by using DVFS and slack reclaiming techniques.

Bibliography

[1] http://www.top500.org/lists/2021/06/.

[2] Yoo A, Jette M, and Grondona Mark. Slurm: Simple linux utility for resource management. In Dror. Rudolph Feitelson and Uwe Larry. Schwiegelshohn, editors, *Job Scheduling Strategies for Parallel Processing*, pages 44–60, Berlin, Heidelberg, 2003. Springer Berlin Heidelberg.

[3] Saeid Abrishami, Mahmoud Naghibzadeh, and Dick Epema. Cost-driven scheduling of grid workflows using partial critical paths. In *2010 11th IEEE/ACM International Conference on Grid Computing*, pages 81–88, 2010.

[4] Saeid Abrishami, Mahmoud Naghibzadeh, and Dick H.J. Epema. Deadline-constrained workflow scheduling algorithms for infrastructure as a service clouds. *Future Generation Computer Systems*, 29(1):158–169, 2013.

[5] Alekh Agarwal and John C. Duchi. Distributed delayed stochastic optimization. In *Proceedings of the 24th International Conference on Neural Information Processing Systems*, NIPS'11, page 873–881, Red Hook, NY, USA, 2011. Curran Associates Inc.

[6] Altair. Pbs professional: Job scheduling andcommercial-grade hpc workload management. Accessed 10 June 2021. http://www.pbsworks.com/Product.aspx?id=1, 2021.

[7] Ei Ando, Toshio Nakata, and Masafumi Yamashita. Approximating the longest path length of a stochastic dag by

a normal distribution in linear time. *Journal of Discrete Algorithms*, 7(4):420–438, 2009.

[8] H. Arabnejad and J.G. Barbosa. List scheduling algorithm for heterogeneous systems by an optimistic cost table. *IEEE Transactions on Parallel and Distributed Systems*, 25(3):682–694, March 2014.

[9] Hamid Arabnejad, Jorge G. Barbosa, and Radu Prodan. Low-time complexity budget-deadline constrained workflow scheduling on heterogeneous resources. *Future Generation Computer Systems*, 55:29–40, 2016.

[10] María Arsuaga-Ríos, Miguel A Vega-Rodríguez, and Francisco Prieto-Castrillo. Meta-schedulers for grid computing based on multi-objective swarm algorithms. *Applied Soft Computing*, 13(4):1567–1582, 2013.

[11] Thet Hsu Aung and Wint Thida Zaw. Improved job scheduling for achieving fairness on apache hadoop yarn. In *2020 International Conference on Advanced Information Technologies (ICAIT)*, pages 188–193, Nov 2020.

[12] Troy Baer, Paul Peltz, Junqi Yin, and Edmon Begoli. Integrating apache spark into pbs-based hpc environments. In *Proceedings of the 2015 XSEDE Conference: Scientific Advancements Enabled by Enhanced Cyberinfrastructure*, XSEDE '15, New York, NY, USA, 2015. Association for Computing Machinery.

[13] S. Bansal, P. Kumar, and K. Singh. An improved duplication strategy for scheduling precedence constrained graphs in multiprocessor systems. *IEEE Transactions on Parallel and Distributed Systems*, 14(6):533 – 544, june 2003.

[14] Savina Bansal, Padam Kumar, and Kuldip Singh. Dealing with heterogeneity through limited duplication for

scheduling precedence constrained task graphs. *Journal of Parallel and Distributed Computing*, 65(4):479 – 491, 2005.

[15] R. Baumann. The impact of technology scaling on soft error rate performance and limits to the efficacy of error correction. In *Digest. International Electron Devices Meeting,*, pages 329–332, 2002.

[16] Anne Benoit, Mourad Hakem, and Yves Robert. Contention awareness and fault-tolerant scheduling for precedence constrained tasks in heterogeneous systems. *parallel computing*, 35(2):83–108, 2009.

[17] Anne Benoit, Mourad Hakem, and Yves Robert. Optimizing the latency of streaming applications under throughput and reliability constraints. In *2009 International Conference on Parallel Processing*, pages 325–332, 2009.

[18] Shishir Bharathi, Ann Chervenak, Ewa Deelman, Gaurang Mehta, Mei-Hui Su, and Karan Vahi. Characterization of scientific workflows. In *2008 Third Workshop on Workflows in Support of Large-Scale Science*, pages 1–10, 2008.

[19] Brett .Bode, David M. Halstead, Ricky .Kendall, Zhou LeI, and David Jackson. The portable batch scheduler and the maui scheduler on linux clusters. In *ALS'00 Proceedings of the 4th annual Linux Showcase & Conference - Volume 4*, pages 27–27, 2000.

[20] Cristina Boeres, Idalmis Milián Sardiña, and Lúcia MA Drummond. An efficient weighted bi-objective scheduling algorithm for heterogeneous systems. *Parallel Computing*, 37(8):349–364, 2011.

[21] Doruk Bozdag, Fusun Ozguner, and Umit V. Catalyurek. Compaction of schedules and a two-stage approach for duplication-based dag scheduling. *IEEE Transactions on Parallel and Distributed Systems*, 20(6):857–871, 2009.

[22] Thomas D Burd and Robert W Brodersen. Energy efficient cmos microprocessor design. In *System Sciences, 1995. Proceedings of the Twenty-Eighth Hawaii International Conference on*, volume 1, pages 288–297. IEEE, 1995.

[23] Louis-Claude Canon and Emmanuel Jeannot. Evaluation and optimization of the robustness of dag schedules in heterogeneous environments. *IEEE Transactions on Parallel and Distributed Systems*, 21(4):532–546, 2010.

[24] Jianguo Chen, Kenli Li, Zhuo Tang, Kashif Bilal, Shui Yu, Chuliang Weng, and Keqin Li. A parallel random forest algorithm for big data in a spark cloud computing environment. *IEEE Transactions on Parallel and Distributed Systems*, 28(4):919–933, 2017.

[25] Juan Chen, Yuhua Tang, Yong Dong, Jingling Xue, Zhiyuan Wang, and Wenhao Zhou. Reducing static energy in supercomputer interconnection networks using topology-aware partitioning. *IEEE Transactions on Computers*, 65(8):2588–2602, 2016.

[26] Weiwei Chen, Rafael Ferreira da Silva, Ewa Deelman, and Thomas Fahringer. Dynamic and fault-tolerant clustering for scientific workflows. *IEEE Transactions on Cloud Computing*, 4(1):49–62, 2016.

[27] K. Christodoulopoulos, V. Sourlas, I. Mpakolas, and E. Varvarigos. A comparison of centralized and distributed meta-scheduling architectures for computation and communication tasks in grid networks. *Computer Communications*, 32(7):1172–1184, 2009.

[28] Charles E. Clark. The greatest of a finite set of random variables. *Operations Research*, 9(2):145–162, 1961.

[29] Sarah Cohen-Boulakia, Khalid Belhajjame, Olivier Collin, Jérôme Chopard, Christine Froidevaux, Alban Gaignard,

Konrad Hinsen, Pierre Larmande, Yvan Le Bras, Frédéric Lemoine, Fabien Mareuil, Hervé Ménager, Christophe Pradal, and Christophe Blanchet. Scientific workflows for computational reproducibility in the life sciences: Status, challenges and opportunities. *Future Generation Computer Systems*, 75:284–298, 2017.

[30] Andrea Colagrossi and Maurizio Landrini. Numerical simulation of interfacial flows by smoothed particle hydrodynamics. *Journal of Computational Physics*, 191(2):448–475, 2003.

[31] M. Cosnard and M. Loi. Automatic task graph generation techniques. In *System Sciences, 1995. Proceedings of the Twenty-Eighth Hawaii International Conference on*, volume 2, pages 113–122. IEEE, 1995.

[32] Y. Cui, E. Poyraz, J. Zhou, S. Callaghan, P. Maechling, T.H. Jordan, L. Shih, and P. Chen. Accelerating cybershake calculations on the xe6/xk7 platform of blue waters. In *2013 Extreme Scaling Workshop (xsw 2013)*, pages 8–17, 2013.

[33] V. JEVTIĆ D. LETIĆ. The distribution of time for clark's flow and risk assessment for the activities of pert network structure, 2009.

[34] Mohammad I. Daoud and Nawwaf Kharma. A high performance algorithm for static task scheduling in heterogeneous distributed computing systems. *Journal of Parallel and Distributed Computing*, 68(4):399 – 409, 2008.

[35] Kalyanmoy Deb, Amrit Pratap, Sameer Agarwal, and TAMT Meyarivan. A fast and elitist multiobjective genetic algorithm: Nsga-ii. *IEEE Transactions on Evolutionary Computation*, 6(2):182–197, 2002.

[36] V. Degalahal, Lin Li, V. Narayanan, M. Kandemir, and M.J. Irwin. Soft errors issues in low-power caches. *IEEE*

Transactions on Very Large Scale Integration (VLSI) Systems, 13(10):1157–1166, 2005.

[37] Ebin Deni Raj, J. P. Nivash, M. Nirmala, and L. D Dhinesh Babu. A scalable cloud computing deployment framework for efficient mapreduce operations using apache yarn. In *International Conference on Information Communication and Embedded Systems (ICICES2014)*, pages 1–6, Feb 2014.

[38] M. L. Dertouzos and A. K. Mok. Multiprocessor online scheduling of hard-real-time tasks. *IEEE Trans. Softw. Eng.*, 15(12):1497–1506, December 1989.

[39] D.V. Djonin, Ashok Karmokar, and V.K. Bhargava. Optimal and suboptimal scheduling over time varying flat fading channels. *IEEE Trans. Wirel. Commun.*, 2:906–910, 07 2004.

[40] A. Dogan and F. Ozguner. Matching and scheduling algorithms for minimizing execution time and failure probability of applications in heterogeneous computing. *IEEE Transactions on Parallel and Distributed Systems*, 13(3):308–323, 2002.

[41] Atakan Doğan and Füsun Özgüner. Biobjective scheduling algorithms for execution time–reliability trade-off in heterogeneous computing systems. *The Computer Journal*, 48(3):300–314, 2005.

[42] Fang Dong, Junzhou Luo, Aibo Song, and Jiahui Jin. Resource load based stochastic dags scheduling mechanism for grid environment. In *2010 IEEE 12th International Conference on High Performance Computing and Communications (HPCC)*, pages 197–204, 2010.

[43] Fangpeng Dong. Scheduling algorithms for grid computing: State of the art and open problems. 2006.

[44] Jack J Dongarra, Emmanuel Jeannot, Erik Saule, and Zhiao Shi. Bi-objective scheduling algorithms for optimizing makespan and reliability on heterogeneous systems. In *Proceedings of the nineteenth annual ACM symposium on Parallel algorithms and architectures*, pages 280–288. ACM, 2007.

[45] M. Dorigo and G. Di Caro. Ant colony optimization: a new meta-heuristic. In *Proceedings of the 1999 Congress on Evolutionary Computation-CEC99 (Cat. No. 99TH8406)*, volume 2, pages 1470–1477 Vol. 2, 1999.

[46] Zhihui Du, Hongyang Sun, Yuxiong He, Yu He, David A. Bader, and Huazhe Zhang. Energy-efficient scheduling for best-effort interactive services to achieve high response quality. In *2013 IEEE 27th International Symposium on Parallel and Distributed Processing*, pages 637–648, 2013.

[47] Juan J. Durillo, Vlad Nae, and Radu Prodan. Multi-objective energy-efficient workflow scheduling using list-based heuristics. *Future Generation Computer Systems*, 36:221–236, 2014.

[48] Juan J Durillo and Antonio J Nebro. jmetal: A java framework for multi-objective optimization. *Advances in Engineering Software*, 42(10):760–771, 2011.

[49] Hesham El-Rewini and T.G. Lewis. Scheduling parallel program tasks onto arbitrary target machines. *Journal of Parallel and Distributed Computing*, 9(2):138–153, 1990.

[50] D. Ernst, S. Das, S. Lee, D. Blaauw, T. Austin, T. Mudge, Nam Sung Kim, and K. Flautner. Razor: circuit-level correction of timing errors for low-power operation. *IEEE Micro*, 24(6):10–20, 2004.

[51] Hamid Mohammadi Fard, Radu Prodan, Juan Jose Durillo Barrionuevo, and Thomas Fahringer. A multi-objective

approach for workflow scheduling in heterogeneous environments. In *Proceedings of the 2012 12th IEEE/ACM International Symposium on Cluster, Cloud and Grid Computing (ccgrid 2012)*, pages 300–309. IEEE Computer Society, 2012.

[52] F. Firouzi, A. Yazdanbakhsh, H. Dorosti, and S. M. Fakhraie. Dynamic soft error hardening via joint body biasing and dynamic voltage scaling. In *2011 14th Euromicro Conference on Digital System Design*, pages 385–392, 2011.

[53] Ian Foster and Carl Kesselman. *The Grid: Blueprint for a New Computing Infrastructure.* 1999.

[54] R.F. Freund and H.J. Siegel. Heterogeneous processing. *IEEE Computer*, 26(6):13 – 17, 1993.

[55] Michael R. Garey and David S. Johnson. *Computers and Intractability: A Guide to the Theory of NP-Completeness.* W. H. Freeman & Co., USA, 1979.

[56] Saurabh Kumar Garg, Chee Shin Yeo, Arun Anandasivam, and Rajkumar Buyya. Environment-conscious scheduling of hpc applications on distributed cloud-oriented data centers. *Journal of Parallel and Distributed Computing*, 71(6):732–749, 2011.

[57] R.L. Graham, E.L. Lawler, J.K. Lenstra, and A.H.G.Rinnooy Kan. Optimization and approximation in deterministic sequencing and scheduling: a survey. In P.L. Hammer, E.L. Johnson, and B.H. Korte, editors, *Discrete Optimization II*, volume 5 of *Annals of Discrete Mathematics*, pages 287–326. Elsevier, 1979.

[58] Yi Gu and Chandu Budati. Energy-aware workflow scheduling and optimization in clouds using bat algorithm. *Future Generation Computer Systems*, 113:106–112, 2020.

[59] T. Hagras and J. Jane brevecek. A high performance, low complexity algorithm for compile-time task scheduling in heterogeneous systems. *Parallel Computing*, 31(7):653 – 670, 2005.

[60] B. Hamidzadeh, Lau Ying Kit, and D.J. Lilja. Dynamic task scheduling using online optimization. *IEEE Transactions on Parallel and Distributed Systems*, 11(11):1151–1163, 2000.

[61] JianJun Han, Man Lin, Dakai Zhu, and Laurence T. Yang. Contention-aware energy management scheme for noc-based multicore real-time systems. *IEEE Transactions on Parallel and Distributed Systems*, 26(3):691–701, 2015.

[62] Robert L. Henderson. Job scheduling under the portable batch system. In Dror G Feitelson and Larry Rudolph, editors, *Job Scheduling Strategies for Parallel Processing*, pages 279–294, Berlin, Heidelberg, 1995. Springer Berlin Heidelberg.

[63] Chih-Chiang Hsu, Kuo-Chan Huang, and Feng-Jian Wang. Online scheduling of workflow applications in grid environments. *Future Generation Computer Systems*, 27(6):860–870, 2011.

[64] Wei Huang, Lingkui Meng, Dongying Zhang, and Wen Zhang. In-memory parallel processing of massive remotely sensed data using an apache spark on hadoop yarn model. *IEEE Journal of Selected Topics in Applied Earth Observations and Remote Sensing*, 10(1):3–19, Jan 2017.

[65] M.A. Iverson, F. Ozguner, and L.C. Potter. Statistical prediction of task execution times through analytic benchmarking for scheduling in a heterogeneous environment. In *Heterogeneous Computing Workshop, 1999.(HCW'99) Proceedings. Eighth*, pages 99–111. IEEE, 1999.

[66] Michael A. Iverson, Füsun Özgüner, and Gregory J. Follen. Parallelizing existing applications in a distributed heterogeneous environment. In *4TH HETEROGENEOUS COMPUTING WORKSHOP*, pages 93–100, 1995.

[67] Emmanuel Jeannot, Erik Saule, and Denis Trystram. Optimizing performance and reliability on heterogeneous parallel systems: Approximation algorithms and heuristics. *Journal of Parallel and Distributed computing*, 72(2):268–280, 2012.

[68] J. P. Jones. Pbs: Portable batch system. *Beowulf Cluster Computing with Windows*, 2001.

[69] Gideon Juve, Ann Chervenak, Ewa Deelman, Shishir Bharathi, Gaurang Mehta, and Karan Vahi. Characterizing and profiling scientific workflows. *Future Generation Computer Systems*, 29(3):682–692, 2013.

[70] Mohammad Kalantari and Mohammad Kazem Akbari. A parallel solution for scheduling of real time applications on grid environments. *Future Generation Computer Systems*, 25(7):704–716, 2009.

[71] Minhaj Ahmad Khan. Scheduling for heterogeneous systems using constrained critical paths. *Parallel Computing*, 38(4):175–193, 2012.

[72] S.J. Kim and J.C. Browne. A general approach to mapping of parallel computation upon multiprocessor architectures. In *International conference on parallel processing*, volume 3, pages 1–8, 1988.

[73] S. Kirkpatrick, C. D. Gelatt, and M. P. Vecchi. Optimization by Simulated Annealing. *Science*, 220(4598):671–680, May 1983.

[74] Yu-Kwong Kwok and I. Ahmad. Dynamic critical-path scheduling: an effective technique for allocating task graphs

to multiprocessors. *IEEE Transactions on Parallel and Distributed Systems*, 7(5):506–521, 1996.

[75] Yu-Kwong Kwok and Ishfaq Ahmad. Static scheduling algorithms for allocating directed task graphs to multiprocessors. *ACM Comput. Surv.*, 31(4):406–471, December 1999.

[76] Kuan-Chou Lai and Chao-Tung Yang. A dominant predecessor duplication scheduling algorithm for heterogeneous systems. *The Journal of Supercomputing*, 44:126–145, 2008.

[77] Y. C. Lee and A. Y. Zomaya. Minimizing energy consumption for precedence-constrained applications using dynamic voltage scaling. In *2009 9th IEEE/ACM International Symposium on Cluster Computing and the Grid*, pages 92–99, May 2009.

[78] Young Choon Lee and Albert Y. Zomaya. Energy conscious scheduling for distributed computing systems under different operating conditions. *IEEE Transactions on Parallel and Distributed Systems*, 22(8):1374–1381, 2011.

[79] Zhang Lei, Gu Tianqi, Zhao Ji, Ji Shijun, Sun Qingzhou, and Hu Ming. An adaptive moving total least squares method for curve fitting. *Measurement*, 49:107–112, 2014.

[80] Jian Li, Tinghuai Ma, Meili Tang, Wenhai Shen, and Yuanfeng Jin. Improved fifo scheduling algorithm based on fuzzy clustering in cloud computing. *Information*, 8(1), 2017.

[81] Kenli Li, Xiaoyong Tang, Bharadwaj Veeravalli, and Keqin Li. Scheduling precedence constrained stochastic tasks on heterogeneous cluster systems. *IEEE Transactions on Computers*, 64(1):191–204, 2015.

[82] Kenli Li, Xiaoyong Tang, and Qifeng Yin. Energy-aware scheduling algorithm for task execution cycles with normal distribution on heterogeneous computing systems. In *2012*

41st International Conference on Parallel Processing, pages 40–47, 2012.

[83] Keqin Li. Scheduling precedence constrained tasks with reduced processor energy on multiprocessor computers. *IEEE Transactions on Computers*, 61(12):1668–1681, 2012.

[84] Keqin Li. Energy and time constrained task scheduling on multiprocessor computers with discrete speed levels. *Journal of Parallel and Distributed Computing*, 95:15–28, 2016.

[85] Keqin Li. Energy-efficient task scheduling on multiple heterogeneous computers: Algorithms, analysis, and performance evaluation. *IEEE Transactions on Sustainable Computing*, 1(1):7–19, 2016.

[86] Keqin Li. Improving multicore server performance and reducing energy consumption by workload dependent dynamic power management. *IEEE Transactions on Cloud Computing*, 4(2):122–137, 2016.

[87] Ping Li, Lei Ju, Zhiping Jia, and Zhiwen Sun. Sla-aware energy-efficient scheduling scheme for hadoop yarn. In *2015 IEEE 17th International Conference on High Performance Computing and Communications, 2015 IEEE 7th International Symposium on Cyberspace Safety and Security, and 2015 IEEE 12th International Conference on Embedded Software and Systems*, pages 623–628, Aug 2015.

[88] Jyh-Han Lin and Jeffrey Scott Vitter. E-approximations with minimum packing constraint violation (extended abstract). In *Proceedings of the Twenty-Fourth Annual ACM Symposium on Theory of Computing*, STOC '92, page 771–782, New York, NY, USA, 1992. Association for Computing Machinery.

[89] C. L. Liu and James W. Layland. Scheduling algorithms for multiprogramming in a hard-real-time environment. *J. ACM*, 20(1):46–61, January 1973.

[90] Zhaobin Liu, Tao Qin, Wenyu Qu, and Weijiang Liu. Dag cluster scheduling algorithm for grid computing. In *2011 14th IEEE International Conference on Computational Science and Engineering*, pages 632–636, 2011.

[91] Miron Livny and Myron Melman. Load balancing in homogeneous broadcast distributed systems. In *Proceedings of the Computer Network Performance Symposium*, page 47–55, New York, NY, USA, 1982. Association for Computing Machinery.

[92] F. Lotfifar and H.S. Shahhoseini. A low-complexity task scheduling algorithm for heterogeneous computing systems. In *Third Asia International Conference on Modelling Simulation, 2009. (AMS '09)*, pages 596 –601, may 2009.

[93] Charng-da Lu. Scalable diskless checkpointing for large parallel systems. 2005.

[94] M. Maheswaran, T. D. Braun, and H. J. Siegel. Heterogeneous distributed computing. *Encyclop edia of Electrical and Electronics Engineering*, 8:679–690, 1999.

[95] Nicole Megow, Marc Uetz, and Tjark Vredeveld. Models and algorithms for stochastic online scheduling. *Math. Oper. Res.*, 31(3):513–525, August 2006.

[96] S. Mittal and J. S. Vetter. A survey of techniques for modeling and improving reliability of computing systems. *IEEE Transactions on Parallel and Distributed Systems*, 27(4):1226–1238, April 2016.

[97] Rolf H. Möhring, Andreas S. Schulz, and Marc Uetz. Approximation in stochastic scheduling: The power of lp-based priority policies. *J. ACM*, 46(6):924–942, November 1999.

[98] Arslan Munir, Sanjay Ranka, and Ann Gordonross. High-performance energy-efficient multicore embedded computing. *IEEE Transactions on Parallel and Distributed Systems*, 23(4):684–700, 2012.

[99] Tooba Nazar, Nadeem Javaid, Moomina Waheed, Aisha Fatima, Hamida Bano, and Nouman Ahmed. Modified shortest job first for load balancing in cloud-fog computing. In Leonard Barolli, Fang-Yie Leu, Tomoya Enokido, and Hsing-Chung Chen, editors, *Advances on Broadband and Wireless Computing, Communication and Applications*, pages 63–76, Cham, 2019. Springer International Publishing.

[100] OLCF. Project website:https://www.olcf.ornl.gov/wp-content/uploads/2015/02/OLCF-User-Group.

[101] Christos H. Papadimitriou and John N. Tsitsiklis. On stochastic scheduling with in-tree precedence constraints. *SIAM Journal on Computing*, 16(1):1–6, 1987.

[102] Michael Pinedo and Gideon Weiss. Scheduling jobs with exponentially distributed processing times and intree precedence constraints on two parallel machines. *Operations Research*, 33(6):1381–1388, 1985.

[103] Karan D. Prajapati, Pushpak Raval, Miren Karamta, and M. Potdar. Comparison of virtual machine scheduling algorithms in cloud computing. *International Journal of Computer Applications*, 83:12–14, 2013.

[104] Xiao Qin and Hong Jiang. A dynamic and reliability-driven scheduling algorithm for parallel real-time jobs executing on heterogeneous clusters. *Journal of Parallel and Distributed Computing*, 65(8):885–900, 2005.

[105] Meikang Qiu and Edwin H. M. Sha. Cost minimization while satisfying hard/soft timing constraints for

heterogeneous embedded systems. *ACM Trans. Des. Autom. Electron. Syst.*, 14(2), April 2009.

[106] Weiwei Qiu, Zibin Zheng, Xinyu Wang, Xiaohu Yang, and Michael R. Lyu. Reliability-based design optimization for cloud migration. *IEEE Transactions on Services Computing*, 7(2):223–236, 2014.

[107] A. Radulescu and A.J.C. van Gemund. Fast and effective task scheduling in heterogeneous systems. In *Proceedings of 9th Heterogeneous Computing Workshop, 2000. (HCW 2000)*, pages 229 –238, 2000.

[108] S. Ranaweera and D.P. Agrawal. A scalable task duplication based scheduling algorithm for heterogeneous systems. In *Proceedings of 2000 International Conference on Parallel Processing*, pages 383 –390, 2000.

[109] Xiaona Ren, Rongheng Lin, and Hua Zou. A dynamic load balancing strategy for cloud computing platform based on exponential smoothing forecast. 09 2011.

[110] Nikzad Babaii Rizvandi, Javid Taheri, and Albert Y. Zomaya. Some observations on optimal frequency selection in dvfs-based energy consumption minimization. *Journal of Parallel and Distributed Computing*, 71(8):1154–1164, 2011.

[111] Maria A. Rodriguez and Rajkumar Buyya. Deadline based resource provisioningand scheduling algorithm for scientific workflows on clouds. *IEEE Transactions on Cloud Computing*, 2(2):222–235, 2014.

[112] Maria A. Rodriguez and Rajkumar Buyya. Budget-driven scheduling of scientific workflows in iaas clouds with fine-grained billing periods. *ACM Trans. Auton. Adapt. Syst.*, 12(2), May 2017.

[113] Daniel Roten, Yifeng Cui, Kim B. Olsen, Steven M. Day, Kyle Withers, William H. Savran, Peng Wang, and Dawei

Mu. High-frequency nonlinear earthquake simulations on petascale heterogeneous supercomputers. In *SC '16: Proceedings of the International Conference for High Performance Computing, Networking, Storage and Analysis*, pages 957–968, 2016.

[114] Michael H. Rothkopf. Scheduling with random service times. *Management Science*, 12(9):707–713, May 1966.

[115] Samuel Russ, Aric Lambert, Joel Camenisch, Vijay Velusamy, Rajesh Rajan, Shailendra Kumar, Rashid Jean-baptiste, Marion Harmon, and Donna Reese. Predictive scheduling for distributed computing. 11 1998.

[116] Stefan Rusu, Simon Tam, Harry Muljono, David Ayers, Jonathan Chang, Raj Varada, Matt Ratta, and Sujal Vora. A 45 nm 8-core enterprise xeon® processor. In *2009 IEEE Asian Solid-State Circuits Conference*, pages 9–12, 2009.

[117] Yassir Samadi, Mostapha Zbakh, and Claude Tadonki. E-heft: Enhancement heterogeneous earliest finish time algorithm for task scheduling based on load balancing in cloud computing. pages 601–609, 07 2018.

[118] Subhash Chander Sarin, Balaj Nagarajan, and Lingrui Liao. *Stochastic scheduling*. Cambridge University Press, Cambridge, 2010. englisch.

[119] Mark Scharbrodt, Thomas Schickinger, and Angelika Steger. A new average case analysis for completion time scheduling. *J. ACM*, 53(1):121–146, January 2006.

[120] G.C. Sih and E.A. Lee. A compile-time scheduling heuristic for interconnection-constrained heterogeneous processor architectures. *IEEE Transactions on Parallel and Distributed Systems*, 4(2):175–187, 1993.

[121] Gilbert C Sih and Edward A Lee. A compile-time scheduling heuristic for interconnection-constrained heterogeneous

processor architectures. *IEEE Transactions on Parallel and Distributed Systems*, 4(2):175–187, 1993.

[122] Nikolay Simakov, Robert A. DeLeon, Martins L. Innus, Matthew D. Jones, Joseph D. White, Steven P. Gallo, Abani M. Patra, and Thomas R. K. Furlani. Slurm simulator: Improving slurm scheduler performance on large hpc systems by utilization of multiple controllers and node sharing. In *Proceedings of the Practice and Experience on Advanced Research Computing*, PEARC '18, New York, NY, USA, 2018.

[123] J. Singh, S. Betha, B. Mangipudi, and N. Auluck. Contention aware energy efficient scheduling on heterogeneous multiprocessors. *IEEE Transactions on Parallel and Distributed Systems*, 26(5):1251–1264, May 2015.

[124] O. Sinnen. *Task scheduling for parallel systems*, volume 60. Wiley-Interscience, 2007.

[125] O. Sinnen, L.A. Sousa, and F.E. Sandnes. Toward a realistic task scheduling model. *IEEE Transactions on Parallel and Distributed Systems*, 17(3):263–275, 2006.

[126] Oliver Sinnen and Leonel Sousa. Communication contention in task scheduling. *IEEE Transactions on Parallel and Distributed Systems*, 16(6):503–515, 2005.

[127] Oliver Sinnen, Andrea To, and Manpreet Kaur. Contention-aware scheduling with task duplication. *Journal of Parallel and Distributed Computing*, 71(1):77–86, 2011.

[128] Martin Skutella and Marc Uetz. Stochastic machine scheduling with precedence constraints. *SIAM Journal on Computing*, 34(4):788–802, 2005.

[129] Seren Soner and Can Özturan. Integer programming based heterogeneous cpu-gpu cluster scheduler for slurm resource

manager. In *2012 IEEE 14th International Conference on High Performance Computing and Communication 2012 IEEE 9th International Conference on Embedded Software and Systems*, pages 418–424, June 2012.

[130] Sanling Song, David Coit, Qianmei W. Feng, and Hao Peng. Reliability analysis for multi-component systems subject to multiple dependent competing failure processes. *IEEE Transactions on Reliability*, 63(1):331–345, 2014.

[131] Vasileios Spiliopoulos, Stefanos Kaxiras, and Georgios Keramidas. Green governors: A framework for continuously adaptive dvfs. In *2011 International Green Computing Conference and Workshops*, pages 1–8, 2011.

[132] I. Stoica, H. Abdel-Wahab, K. Jeffay, S.K. Baruah, J.E. Gehrke, and C.G. Plaxton. A proportional share resource allocation algorithm for real-time, time-shared systems. In *17th IEEE Real-Time Systems Symposium*, pages 288–299, 1996.

[133] AKM Talukder, Michael Kirley, and Rajkumar Buyya. Multiobjective differential evolution for scheduling workflow applications on global grids. *Concurrency and Computation: Practice and Experience*, 21(13):1742–1756, 2009.

[134] Xiaoyong Tang, Kenli Li, Renfa Li, and Bharadwaj Veeravalli. Reliability-aware scheduling strategy for heterogeneous distributed computing systems. *Journal of Parallel and Distributed Computing*, 70(9):941–952, 2010.

[135] Xiaoyong Tang, Kenli Li, and Guiping Liao. An effective reliability-driven technique of allocating tasks on heterogeneous cluster systems. *Cluster Computing*, 17(4):1413–1425, 2014.

[136] Xiaoyong Tang, Kenli Li, Guiping Liao, Kui Fang, and Fan Wu. A stochastic scheduling algorithm for precedence

constrained tasks on grid. *Future Generation Computer Systems*, 27(8):1083–1091, 2011.

[137] Xiaoyong Tang, Kenli Li, Guiping Liao, and Renfa Li. List scheduling with duplication for heterogeneous computing systems. *Journal of Parallel and Distributed Computing*, 70(4):323–329, 2010.

[138] Xiaoyong Tang, Kenli Li, Meikang Qiu, and Edwin H-M Sha. A hierarchical reliability-driven scheduling algorithm in grid systems. *J. Parallel Distrib. Comput.*, 72(4):525–535, 2012.

[139] Xiaoyong Tang, Kenli Li, Zeng Zeng, and Bharadwaj Veeravalli. A novel security-driven scheduling algorithm for precedence-constrained tasks in heterogeneous distributed systems. *IEEE Transactions on Computers*, 60(7):1017–1029, 2011.

[140] Xiaoyong Tang, Xiaochun Li, and Zhuojun Fu. Budget-constraint stochastic task scheduling on heterogeneous cloud systems. *Concurrency and Computation: Practice and Experience*, 29(19):e4210, 2017.

[141] Xiaoyong Tang, Weiqiang Shi, and Fan Wu. Interconnection network energy-aware workflow scheduling algorithm on heterogeneous systems. *IEEE Transactions on Industrial Informatics*, 16(12):7637–7645, 2020.

[142] Xiaoyong Tang and Tan Weizhen. Energy efficient reliability-aware scheduling algorithm on heterogeneous systems. *Scientific Programming*, page 9823213, March 2016.

[143] OpenPBS Team. A batching queuing system. Accessed 10 June 2021. http://www.openpbs.org/, 2021.

[144] Van Tilborg and Wittie. Wave scheduling—decentralized scheduling of task forces in multicomputers. *IEEE Transactions on Computers*, C-33(9):835–844, 1984.

[145] S. Tongsima, E.H.-M. Sha, C. Chantrapornchai, D.R. Surma, and N.L. Passos. Probabilistic loop scheduling for applications with uncertain execution time. *IEEE Transactions on Computers*, 49(1):65–80, 2000.

[146] H. Topcuoglu, S. Hariri, and Min-You Wu. Performance-effective and low-complexity task scheduling for heterogeneous computing. *IEEE Transactions on Parallel and Distributed Systems*, 13(3):260–274, 2002.

[147] Nick Trebon and Pete Beckman. Empirical-based probabilistic upper bounds for urgent computing applications. In *2008 IEEE International Conference on Cluster Computing*, pages 342–347, 2008.

[148] Tatsuhiro Tsuchiya, Tetsuya Osada, and Tohru Kikuno. Genetics-based multiprocessor scheduling using task duplication. *Microprocessors and Microsystems*, 22(3):197–207, 1998.

[149] Tatsuhiro Tsuchiya, Tetsuya Osada, and Tohru Kikuno. Genetics-based multiprocessor scheduling using task duplication. *Microprocessors and Microsystems*, 22(3):197–207, 1998.

[150] Mueen Uddin, Yasaman Darabidarabkhani, Asadullah Shah, and Jamshed Memon. Evaluating power efficient algorithms for efficiency and carbon emissions in cloud data centers: A review. *Renewable and Sustainable Energy Reviews*, 51:1553–1563, 2015.

[151] J. D. Ullman. Np-complete scheduling problems. *J. Comput. Syst. Sci.*, 10:384–393, June 1975.

[152] Vinod Kumar Vavilapalli, Arun C. Murthy, Chris Douglas, Sharad Agarwal, Mahadev Konar, Robert Evans, Thomas Graves, Jason Lowe, Hitesh Shah, Siddharth Seth, Bikas Saha, Carlo Curino, Owen O'Malley, Sanjay Radia, Benjamin Reed, and Eric Baldeschwieler. Apache hadoop yarn: Yet another resource negotiator. In *Proceedings of the 4th Annual Symposium on Cloud Computing*, SOCC '13, New York, NY, USA, 2013.

[153] Vasanth Venkatachalam and Michael Franz. Power reduction techniques for microprocessor systems. *ACM Comput. Surv.*, 37(3):195–237, September 2005.

[154] Xiaofeng Wang, Chee Shin Yeo, Rajkumar Buyya, and Jinshu Su. Optimizing the makespan and reliability for workflow applications with reputation and a look-ahead genetic algorithm. *Future Generation Computer Systems*, 27(8):1124–1134, 2011.

[155] Yan Wang, Kenli Li, Hao Chen, Ligang He, and Keqin Li. Energy-aware data allocation and task scheduling on heterogeneous multiprocessor systems with time constraints. *IEEE Transactions on Emerging Topics in Computing*, 2(2):134–148, 2014.

[156] Marek Wieczorek, Andreas Hoheisel, and Radu Prodan. Towards a general model of the multi-criteria workflow scheduling on the grid. *Future Generation Computer Systems*, 25(3):237–256, 2009.

[157] Di Wu, Qingxi Hu, Yuan Yao, Gaochun Xu, and Minglun Fang. A study on workflow technology for rp application. In *2009 WRI World Congress on Computer Science and Information Engineering*, volume 3, pages 1–5, 2009.

[158] M.-Y. Wu and D.D. Gajski. Hypertool: a programming aid for message-passing systems. *IEEE Transactions on Parallel and Distributed Systems*, 1(3):330–343, 1990.

[159] Bin Xiang, Bibo Zhang, and Lin Zhang. Greedy-ant: Ant colony system-inspired workflow scheduling for heterogeneous computing. *IEEE Access*, 5:11404–11412, 2017.

[160] Liu Xiaoqian, Yang Shoubao, Wang Shuling, and Xu Jing. R-dls: An improved dls algorithm. In *2008 Seventh International Conference on Grid and Cooperative Computing*, pages 97–101, 2008.

[161] Guoqi Xie, Gang Zeng, Liangjiao Liu, Renfa Li, and Keqin Li. Mixed real-time scheduling of multiple dags-based applications on heterogeneous multi-core processors. *Microprocessors and Microsystems*, 47:93–103, 2016.

[162] Min Xie, Yutong Lu, Kefei Wang, Lu Liu, Hongjia Cao, and xuejun yang. Tianhe-1a interconnect and message-passing services. *IEEE Micro*, 32(1):8–20, 2012.

[163] Chuanfu Xu, Xiaogang Deng, Lilun Zhang, Jianbin Fang, Guangxue Wang, Yi Jiang, Wei Cao, Yonggang Che, Yongxian Wang, Zhenghua Wang, Wei Liu, and Xinghua Cheng. Collaborating cpu and gpu for large-scale high-order cfd simulations with complex grids on the tianhe-1a supercomputer. *Journal of Computational Physics*, 278:275–297, 2014.

[164] M.Q. Xu. Effective metacomputing using lsf multicluster. In *Proceedings First IEEE/ACM International Symposium on Cluster Computing and the Grid*, pages 100–105, 2001.

[165] Yuming Xu, Kenli Li, Ligang He, and Tung Khac Truong. A dag scheduling scheme on heterogeneous computing systems using double molecular structure-based chemical reaction optimization. *Journal of Parallel and Distributed Computing*, 73(9):1306–1322, 2013.

[166] Yuming Xu, Kenli Li, Ligang He, Longxin Zhang, and Keqin Li. A hybrid chemical reaction optimization scheme

for task scheduling on heterogeneous computing systems. *IEEE Transactions on Parallel and Distributed Systems*, 26(12):3208–3222, 2015.

[167] Shiueng-Bien Yang and Shen-I Yang. New decision tree based on genetic algorithm. In *2010 International Symposium on Computer, Communication, Control and Automation (3CA)*, volume 1, pages 115–118, 2010.

[168] Tao Yang and A. Gerasoulis. Dsc: scheduling parallel tasks on an unbounded number of processors. *IEEE Transactions on Parallel and Distributed Systems*, 5(9):951–967, 1994.

[169] Longxin Zhang, Kenli Li, Changyun Li, and Keqin Li. Bi-objective workflow scheduling of the energy consumption and reliability in heterogeneous computing systems. *Information Sciences*, 379:241–256, 2017.

[170] Longxin Zhang, Kenli Li, Keqin Li, and Yuming Xu. Joint optimization of energy efficiency and system reliability for precedence constrained tasks in heterogeneous systems. *International Journal of Electrical Power & Energy Systems*, 78:499–512, 2016.

[171] Longxin Zhang, Kenli Li, Yuming Xu, Jing Mei, Fan Zhang, and Keqin Li. Maximizing reliability with energy conservation for parallel task scheduling in a heterogeneous cluster. *Information Sciences*, 319:113–131, 2015.

[172] Longxin Zhang, Kenli Li, Weihua Zheng, and Kenqin Li. Contention-aware reliability efficient scheduling on heterogeneous computing systems. *IEEE Transactions on Sustainable Computing*, 3(3):182–194, 2018.

[173] Baoxian Zhao, Hakan Aydin, and Dakai Zhu. Shared Recovery for Energy Efficiency and Reliability Enhancements in Real-Time Applications with Precedence Constraints. *ACM*

Transactions on Design Automation of Electronic Systems, 18(2), MAR 2013.

[174] Henan Zhao and Rizos Sakellariou. An experimental investigation into the rank function of the heterogeneous earliest finish time scheduling algorithm. In Harald Kosch, , László Böszörményi, and Hermann Hellwagner, editors, *Euro-Par 2003 Parallel Processing*, pages 189–194, Berlin, Heidelberg, 2003. Springer Berlin Heidelberg.

[175] Laiping Zhao, Yizhi Ren, and Kouichi Sakurai. Reliable workflow scheduling with less resource redundancy. *Parallel Computing*, 39(10):567–585, 2013.

[176] Laiping Zhao, Yizhi Ren, Yang Xiang, and Kouichi Sakurai. Fault-tolerant scheduling with dynamic number of replicas in heterogeneous systems. In *2010 IEEE 12th International Conference on High Performance Computing and Communications (HPCC)*, pages 434–441, 2010.

[177] Zibin Zheng, Tom Chao Zhou, Michael R. Lyu, and Irwin King. Component ranking for fault-tolerant cloud applications. *IEEE Transactions on Services Computing*, 5(4):540–550, 2012.

[178] D. Zhu, R. Melhem, and D. Mosse. The effects of energy management on reliability in real-time embedded systems. In *IEEE/ACM International Conference on Computer Aided Design, 2004. ICCAD-2004.*, pages 35–40, 2004.

[179] Dakai Zhu and Hakan Aydin. Reliability-aware energy management for periodic real-time tasks. In *13th IEEE Real Time and Embedded Technology and Applications Symposium (RTAS'07)*, pages 225–235, 2007.

[180] Dakai Zhu, Rami Melhem, and Daniel Mossé. The effects of energy management on reliability in real-time

embedded systems. In *Computer Aided Design, 2004. ICCAD-2004. IEEE/ACM International Conference on,* pages 35–40. IEEE, 2004.

[181] Xiaomin Zhu, Rong Ge, Jinguang Sun, and Chuan He. 3e: Energy-efficient elastic scheduling for independent tasks in heterogeneous computing systems. *Journal of Systems and Software,* 86(2):302–314, 2013.

[182] A.Y. Zomaya, C. Ward, and B. Macey. Genetic scheduling for parallel processor systems: comparative studies and performance issues. *IEEE Transactions on Parallel and Distributed Systems,* 10(8):795–812, 1999.

9781032309200